OPTIMAL CONTROL FOR ECONOMETRIC MODELS

OPTIMAL CONTROL FOR ECONOMETRIC MODELS

An Approach to Economic Policy Formulation

edited by

Sean Holly, Berç Rüstem and Martin B. Zarrop

M

First published 1979 by
THE MACMILLAN PRESS LTD
London and Basingstoke
Associated companies in Delhi
Dublin Hong Kong Johannesburg Lagos
Melbourne New York Singapore Tokyo

Printed in Great Britain by
J. W. Arrowsmith Ltd, Bristol BS3 2NT

British Library Cataloguing in Publication Data

Recent Developments in National Economic Policy Formulation (*Conference*), *London, 1977*
Optimal control for econometric models
1. Economics – Mathematical models – Congresses
2. Econometrics – Methodology – Congresses
3. Control theory – Congresses
I. Title II. Holly, Sean III. Rüstem, Berç
IV. Zarrop, Martin B. V. Imperial College of Science and Technology. Department of computing and control
330'.01'82 HB141

ISBN 0–333–25706–5

Contents

PART IV STABILISATION POLICIES WITH LARGE NONLINEAR ECONOMETRIC MODELS

Foreword

In view of the interest that has recently been shown in national policy formulation it was thought useful in the spring of 1977 to hold a conference on 'Recent Developments in National Economic Policy Formulation'. The idea was to provide a forum to practising economists in academic and Government circles at which ideas and recent developments in economic policy formulation through the use of macroeconomic models could be presented and discussed.

In the event just over 100 invited participants took part in the conference, which was held at Imperial College on 19–21 April 1977 under the auspices of the Department of Computing and Control. Financial assistance from the Social Science Research Council enabled a number of distinguished contributors from overseas to attend and present papers at the conference and to enliven and broaden the basis of the discussion. The choice of date, being about a month after the UK Budget announcement, had the happy outcome of attracting a pleasingly high attendance of UK Treasury officials, to the benefit of the free-ranging discussions that occurred.

It is now felt that the analytical techniques and technical material that were presented at the conference deserves closer study by a wider audience; and hence this volume, whose chapters are based on papers presented then.

I would like to record our thanks to members of the Steering Committee for the conference for their good advice that enabled us to get together such a lively and informed conference, to the authors for their interesting and learned papers and to the editors of this volume who have tried to distil the quintessence of the proceedings in the twelve chapters of this book.

J. H. Westcott

Programme of Research in Econometric Methods
Department of Computing and Control
Imperial College

Preface

This volume is based on a conference on 'Recent Developments in Economic Policy Formulation' held in April 1977 by the SSRC-financed Programme of Research into Econometric Methods in the Department of Computing and Control, Imperial College London.

A few of the papers have been altered for presentation in this volume. Terry Barker (Chapter 12) substantially revised his paper while Mark Salmon and Peter Young (Chapter 6) prepared their contribution specially for this volume. The paper by Westcott, Zarrop, Holly, Rüstem and Becker (Chapter 1) has been largely rewritten so as to provide a general introductory essay on optimal control in economic policy-making.

The Programme of Research into Econometric Methods was originally set up under the joint direction of Dr J. Bray, Professor M. Peston and Professor J. H. Westcott in 1971. A debt of gratitude is owed to P. E. Caines, K. D. Wall and A. J. Preston, who carried out much of the early theoretical work, as well as to Phillipa Carling, Sally Donovan, M. Kharusi, J. Peddicord, M. Salmon and K. Vellupillai, without whose efforts the present work of the Programme would have been impossible.

We wish to thank Helen Rogers for her patience and typing skill during the preparation for the conference and in preparing a final typescript for publication.

SEAN HOLLY
BERÇ RÜSTEM
MARTIN B. ZARROP

List of Contributors

Dr T. S. Barker	Cambridge University
Mr R. G. Becker	Imperial College London
Professor G. C. Chow	Princeton University
Professor J. F. Helliwell	University of British Columbia
Dr S. Holly	Imperial College London
Dr A. Hughes-Hallett	Erasmus University, Rotterdam
Professor D. Kendrick	University of Texas
Professor L. R. Klein	University of Pennsylvania
Dr D. A. Livesey	Cambridge University
Dr M. J. O'Connell	Science Research Council (Rutherford Laboratory)
Mr B. Rüstem	London School of Economics
Mr M. Salmon	Australian National University, Canberra
Professor J. H. Westcott	Imperial College London
Professor P. Young	Australian National University, Canberra
Dr M. B. Zarrop	Imperial College London

Introduction

The history of attempts to introduce control theoretic concepts into the field of economics goes back at least twenty years. It has only been in the last decade, however, that the explosion in computer technology coupled with increasing sophistication in econometric modelling have opened the way to rapid progress in the formulation and solving of optimal stabilisation problems. Control theory has also been of use in the theoretical investigation of dynamic economic systems. The papers in this volume belong primarily to the first and more applied strand in the literature and are addressed mainly to aspects of the application of control methods to macroeconomic policy-making.

The contents have been divided into four parts so as to bring loosely related papers under common headings. The first and introductory paper by Westcott *et al.* is designed to convey the salient features of optimal economic policy-making. It is far from being an exhaustive account since to do justice to the entire field would require a book in itself. The paper covers many of the important concepts that arise repeatedly in subsequent chapters and places optimal policy-making and current practices in the broader context of constrained optimisation.

Part II contains a number of differing papers. Chow describes a twelve-step procedure for applying stochastic optimal control techniques to econometric models. A means of judging between the possibly contradictory recommendations of more than one model is proposed in terms of a payoff matrix. Minimising the expected loss associated with alternative policies can then be achieved by minimax strategies or by assigning probabilities to the different models and adopting a Bayesian decision procedure. The contribution by Helliwell is concerned with drawing out the implications for optimal policy-making of recently developed theories of open economies under fixed and flexible exchange rates. The papers by Livesey and by Salmon and Young are similar in approach in so far as they seek to derive their inspiration, not from 'modern' optimal control theory, which relies very heavily upon state space representations, but from recent developments and improvements in 'classical' theory. Particular emphasis is placed upon the concepts of stability and robustness rather than on the more narrow notion of optimality. Both these papers provide a contrast (and an antidote, perhaps?) to the first paper by Westcott *et al.*

The papers by Rüstem *et al.* and by Hughes-Hallett are the most technically demanding in the volume. The Rüstem *et al.* paper examines the problem of how to obtain plausible trajectories for targets and instruments by using a quadratic objective function given the difficulties that confront a policy-maker if he is asked to attach numerical values to marginal deviations of economic targets from desired paths. A solution is provided in the form of an algorithm that respecifies the weights in the quadratic objective function. The policy-maker is only asked to make his comments on the 'optimal' trajectories for each of the relevant economic variables and – if he finds any of these unacceptable – to indicate preferred values. The essence of the method by which these preferred values are then translated back into weights in the objective function consists in making a correction to the current weighting matrix which will then push the next optimal solution closer to the preferred solution. The correction employs an updating formula widely used in computing approximations to second derivative matrices in unconstrained and constrained optimisation. Hughes-Hallett, in his paper, considers the sensitivity of the paths of both targets and instruments to perturbations in the parameters of the econometric model and to perturbations in the weights in the objective function. Some numerical results for Klein's model indicate that the optimal paths are more sensitive to variations in the model parameters than in the weights in the objective function.

Part III contains two papers on optimal control of linear models. The Holly *et al.* paper uses a linear model of the UK economy to illustrate the sensitivity of the optimal target paths to variations in the paths for the exogenous variables. Using time-varying weights in the objective function, constraints on instrument usage are incorporated into the solution. The paper by Kendrick reports some preliminary results of a study of the impact of measurement error on economic policies, and the contribution that adaptive control, which takes account of the possibility that model parameters can be learned, can make to minimising their impact.

Economic policy-making is normally conducted with the assistance of a large nonlinear econometric model but much of the established optimal control theory pertains only to linear systems. The three papers of Part IV are addressed to various aspects of the problem of using a nonlinear model to generate economic policies. The first paper, by Zarrop *et al.*, approaches the problem from an unusual point of view.

The basic idea is to control the nonlinear model, in this case the London Business School model, by applying a feedback law which is optimal for some linear model, in this case the model described in Chapter 8.

The contribution by Klein has been included in the section on large models, though some shortcomings of present demand-orientated econometric models, data inadequacies and modelling errors are also examined. The section on optimal policy-making describes some results obtained with the Wharton Model using direct optimisation techniques. Details of these techniques are given in Section 14 of Chapter 1. The final paper, by Barker, describes a means of obtaining economic policies from the dynamic input-output model of the Cambridge Growth Project. The absence of an objective function makes the analysis incomplete from the viewpoint of an optimal solution, but it is one of the first attempts to tackle the difficulties of policy formulation with dynamic input-output models.

PART I

1 A Control Theory Framework for Policy Analysis

J. H. Westcott, M. B. Zarrop, S. Holly, B. Rüstem and R. Becker

1. INTRODUCTION

In the last decade there have been major changes in the manner in which governments have decided upon what kinds of economic policies to pursue. Large econometric models which provide forecasts and information about possible policy alternatives are now a common feature. The degree to which they are used is not uniform among countries and even where considerable reliance is placed upon them very often other considerations will predominate. More recently attention has shifted to developing ways in which these models could be used more effectively. One area that has attracted widespread attention is that of stochastic optimal control theory.

There is always the danger, however, with the introduction of new techniques, that they will be regarded—by their advocates at least—as the panacea for all economic ills. When they fail to live up to this inflated expectation, as has been the case with econometric forecasting models, the inevitable reaction tends to distract attention from the undoubted gains. The control approach does not provide a ready-made set of tools that will allow us to calculate the values of economic instruments for the next five years and then sit back and wait for the results.

An attempt to reduce the complex processes of political decision-making to a relatively straightforward exercise in constrained optimisation may appear impossibly ambitious. Although economists often draw a distinction between 'normative' and 'positive' considerations it does not always seem possible to disentangle the two levels of discourse.

A disagreement about the proper ends of policy can easily become confused with a disagreement about the alternative means and their quantitative significance. To compound the confusion, there may be opinions expressed about the choice between means as well as between ends. These sorts of difficulties are added to by the tendency for political discussion to blur the often unpleasant choices at the margin between different groups differentially affected by different policy combinations. These points raise the question whether in reality we can formulate the decision-making process in terms of some quadratic, or even non-quadratic, welfare or objective function. Control theory cannot make any contribution to the resolution of irreconcilable differences of opinion in political life. Nevertheless, we can observe that even in the face of the agony of indecision, decisions are made, things do get done; and it is with the quality of these decisions that we are concerned.

The paper is divided into two main parts. The first gives a reasonably broad, descriptive exposition of control theory as it has been applied to macroeconomic policy-making at least in the academic literature. This should provide an introduction to the main concepts for those who find them unfamiliar. The second part is more technical and covers briefly the state space approach to the representation of econometric models, the linear, quadratic, gaussian approach to optimal control and direct constrained optimisation of nonlinear models.

PART I

2. MODELLING FOR FORECASTING

One of the basic developments since the war in the application of economics to policy-making has been the quantification of fundamental economic relationships, and their bringing together in the form of a model that describes the workings of the economy as a system of linear and nonlinear equations.

A primary requirement for any control exercise is a model that describes the general relationship between economic instruments (inputs, controls) and endogenous variables (outputs). It is common in engineering to model the system using what is called the 'black box' approach, whereby the direct relationship between inputs and outputs is considered and the parameters of the dynamic process estimated without prior structural assumptions. In the modelling of economic systems, however, economists generally are reluctant to adopt this approach.

They have a preference for structural models. The mechanism by which an instrument transmits its effects to an output is masked when a reduced form or input-output mapping is used. In the structural form the causal links by which an instrument works its way from sector to sector through the economy are spelt out as clearly as the data will permit. In macroeconomics there is an absence of any sufficiently well grounded and widely accepted body of comprehensive theory, such as the engineer finds in the laws of physics, that can give the economic modeller more confidence in his representation of the economy. The uniqueness of historical data, as well as its paucity, is another major problem. Since controlled experimentation is impossible the economist must rely upon a single set of data which is subject to distressingly frequent revision and changes in definition. The effect of all this is to make the modeller reluctant to accept the model and the results it produces at face value.

The gradual development of econometric models has been one of organic adaptation and growth. At first only a relatively simple structure was erected, based on the system of national accounts. With time the models have been expanded by the addition of new sectors and the disaggregation of existing relationships. As experience has accumulated, especially in forecasting and policy simulations, various shortcomings of the existing model have been detected and this in turn has led to further modification. Forecasting provides a means of evaluating the performance of a model but it is not the only criterion. A model under simulation is expected to have various long-run characteristics that are suggested by economic theory, and to be able to track reasonably well the behaviour of the economy during the estimation period.

Short-term forecasting now plays a central role in present methods of demand management. Forecasts of what would actually happen over the next 18 months are made on the assumption that there is no change in policy. Forecasts of the consequences of alternative policies are then made, and it is on the basis of these policy simulations that the Government decides what to do. This is not a mechanical exercise involving only a computerised model. There will always be a degree of uncertainty about the performance of particular equations of the model and their forecasts may not be consistent with qualitative information or with forecasts obtained using survey methods. Furthermore, the forecaster may have information about unsystematic factors affecting the equations that are rightly included in the error term in the estimation stage of model building. The consequence of a strike or a drought during a particular time period can be accommodated by careful adjustment of the intercept of an equation. Thus, forecasting accuracy requires a very detailed econometric model and the systematic inclusion of other information about the disturbances affecting each equation of the model. Needless to say, information about disturbances entering the system is likely to be confined to at most the first few quarters of the forecasting period. There may, however, be systematic effects, such as the advent of North Sea oil, which have an impact over a considerable period into the future and these must be taken into account also.

In quite a short space of time considerable international experience has been gained of the pitfalls and difficulties of short-term forecasting with econometric models. Earlier optimism about the capacity of the models to improve forecasting has given way to a considerable degree of scepticism about the ability of forecasters, whether they are using formal or informal methods, to produce sufficiently accurate short-term forecasts that can act as a guide to economic management. It is possible that we have been asking too much from our economic models. The partial nature of economic theory as an explanation of behaviour and the roundabout and inferential character of our knowledge of the workings of the economy, coupled to the stochastic nature of macroeconomic events, suggest that the predictive power of our models is inherently low. No matter how detailed the models, there are certain natural limits to the accuracy of forecasts determined by the standard errors of equations. The possibility of eventually reducing the forecasting error to some irreducible minimum is ruled out by the adaptive character of the economic system and the occurrence of structural change. Rather than frame economic policies so that their success or failure depends on the accuracy of economic forecasting, it is better that policies should reflect the inevitability of forecasting error and be framed accordingly. It is here that control theory may be able to make some contribution.

3. SOME ASPECTS OF CONTROL AND OPTIMALITY

In the engineering field, control theory is concerned with achieving desired system behaviour through manipulation of the plant inputs. This is to be carried out without violating given physical constraints and in the face of uncertainties and disturbances. This is also the position in which the policy-maker finds himself. The nature of the economic process leads naturally to its description in stochastic rather than in deterministic terms on a macroeconomic level. However detailed and complex the model, unexpected events having economic repercussions but arising outside the chosen level of description must be taken into account. Thus forecasting is always endowed with some uncertainty (as in the real world) and this fact of life is reflected in the choice of a stochastic model. Given the stochastic nature of the economy, policy optimisation can only be concerned with controlling the controllable, i.e. guiding the underlying economic process in a desirable direction and in such a way that the effects of random shocks are minimised. Control theory is concerned precisely with such problems.

Regulation of an engineering system, in the sense of maintaining some desired steady-state regime in an uncertain environment, is the bread-and-butter of the control engineer. Long experience has shown that the nature of the relationship which must be established between the outputs and the controls to achieve satisfaction may be extremely simple in form, despite the complexity of the plant itself. There exist a number of important policy problems in economics that can be posed—at least as a first approximation—as regulator problems in the engineering sense.

The success of control applications in the engineering field has caught the attention of many economists. As long ago as the 1950s Phillips, among others, recognised the importance of control theory for the regulation of economic systems. It is only in the last decade, however, that the necessary maturity has been achieved in econometric model building and control theory, and the requisite computing technology become widely available. Nevertheless, many economists involved in the process of policy formulation remain suspicious of what appears a rarefied approach to practical matters. These suspicions have been reinforced by the frequency with which 'optimal' is used in conjunction with the term 'control theory'.

Where the measure of acceptability of a policy is embodied in a welfare function, the term 'optimality' can be given a precise meaning. An optimal policy is one that minimises the welfare function without violating any chosen constraints. Thus optimal implies neither 'good' nor 'bad' but may simply be a way of selecting one policy strategy from a number of satisfactory scenarios.

In this respect the situation facing both control engineers and economists is the same. To quote a member in the first category: 'There is no such thing as an optimal design process, since it depends very strongly on the specific application, design experience and time available to do the job' (Athans (1971)). Thus, the hard work has to be carried out at the stages of modelling and in choosing the welfare function, i.e. in setting up the optimisation problem, and is in no way obviated by waving a magic wand labelled 'optimality'. Adequate experience is sadly lacking in this area and a great deal of further work is needed.

Once the optimisation problem is posed in a satisfactory way, its solution may be difficult because of the size and nonlinearity of the economic model. Various simplifications are possible but, in the strictly mathematical sense, what is optimal in a simplified scheme will not be, in general, an optimal solution to the original problem. However, bearing in mind our remarks on the conditional nature of optimality, the derived solution may well be completely satisfactory to the policy planner. Therefore the main point that must be stressed about the use of optimal control theory in economic policy formulation is not that it disgorges some unique panacea but that it provides a consistent and rational framework for processing and integrating into a policy strategy over a given period the policy-maker's objectives and constraints while taking into account the dynamics of the economic system.

There are basically two approaches to solving a constrained optimisation problem with only equality constraints. The first is the well-known Lagrange multiplier technique absorbing the constraints into a new objective function for which the unconstrained optimum is sought. Chow (1975) derives all the relations for dynamic programming in connection with the solution of the LQG problem (Linear Model–Quadratic Objective Function–Gaussian Disturbance) using this approach. The second approach is to eliminate the equality constraints by substituting them directly into the objective function. The solution to the original constrained optimisation problem is also the unconstrained optimum of the transformed objective function. This is the basis of the approaches used by Fair (1974), Holbrook (1974), Athans *et al.* (1976) and Preston *et al.* (1976) although the form of their respective objective functions, models and other constraints is different. For instance, Fair substitutes his nonlinear model into a nonlinear and nonquadratic objective function. Optimal control problems can also be brought within this framework following the methods described in Canon *et al.* (1971) and Polak (1971).

For nonlinear models the linearisation of the model equations is sometimes carried out before substitution into the objective function. For example, Holbrook linearises the nonlinear Michigan Model of the US economy and then substitutes the linearisation into a quadratic objective function. The linearisation may be repeated at each iteration of the algorithm. If convergence is assumed, then the values of the endogenous variables computed by the model equal those computed by its linear approximation.

The optimum generated by a constrained optimisation problem is generally termed an open-loop solution in control theory. The reason for this is that the values of the policy instruments at some instant of time do not, in general, depend on the actual performance of the system up to and including that time. This is inadequate in the presence of random disturbances to the system when future behaviour is uncertain. However, in the specific case when the optimal policy instruments can be analytically expressed as functional relationships of the current and past values of the system variables, the solution is said to be a closed-loop feedback solution. The functional relationships involved are given by the optimal solution itself. This type of solution is discussed further in Appendix B.

In a perfectly known and disturbance-free world there would be no advantage in employing our chosen policy other than as a sequence of fixed and predetermined values for the relevant instruments. In such a world one could drive blindfold from London to Brighton by using precomputed control actions. Such a solution would be open-loop. Unfortunately, in the real world it would quickly lead to disaster. A closed-loop feedback solution is one in which a control law is established, i.e. a functional relationship between current policy and current and past values of the economic variables, so that there is a feedback of information from the environment. This would imply, in economic parlance, the introduction of an automatic stabiliser. In this case deviations of endogenous variables from expected values due to stochastic effects change the values of the policy instruments and introduce a self-correcting behaviour. Indeed, such a feedback law changes the dynamics of the economic model.

4. THE ROLE OF THE OBJECTIVE FUNCTION

A number of arguments have been raised against the use of formalised objective functions in economic policy-making. Some are directed at all kinds of function, while others are critical of the specific use of quadratic objective functions. A clear idea of the role of the objective function can be gained from regarding it in the context of constrained optimisation discussed in the previous section. The purpose of using any convex objective function in an optimisation framework is to generate a unique set of model variables, both targets and instruments. However, these values must have the primary property of being feasible by satisfying the model equations and any other imposed constraints. Optimality with respect to the chosen objective function is a secondary quality, but is important in selecting a particular set of variable values.

The problem facing the policy-maker, once the purely exogenous variables in his model have been set, is the determination of satisfactory values for the policy instruments and the outputs. Current practice is based on repeated simulations of the economy using a large macroeconomic model, and the essentially iterative procedure can be summed up in three basic steps involving interaction between the 'policy-maker' and a 'modeller':

Step 1: Given values for the instruments, the modeller computes the output values.
Step 2: The policy-maker evaluates the economic relevance of the instruments and output values. If they are satisfactory, the process terminates. If not, specific criticisms are made to the modeller.
Step 3: The modeller proposes corrections to the instrument values in order to bring the output values into line with the policy-maker's requirements. The corrections are based on expectations of model behaviour, grounded in experience of many previous simulations as well as on the intuition and judgement of the modeller.

All existing methods for rationalising the decision-making process can be interpreted as special cases within the general framework of constrained optimisation theory. The constraints are primarily the model equations together with any other equality and inequality constraints.

The general decision-making process can be formalised as follows.

Step A: Generate the optimal values of the instruments and outputs by minimising an objective function subject to the relevant constraints;
Step B: Evaluation phase (same as Step 2 above);
Step C: The modeller respecifies the optimisation problem and returns to Step A.

Note that, in contrast to Step 1, Step A itself will normally employ an iterative algorithm.

It is useful to examine the process of economic policy formulation via simulation in this context. The simulation procedure can be interpreted as minimising some function of the instruments alone subject to the model constraints. Such an objective function is particularly simple. However, it is certain that any realistic objective function actually expressing the policy-maker's goals will be more general in structure and involve the outputs as well.

A further argument for a more general objective function arises from the optimisation context itself. The number of degrees of freedom is equal to the total number of instruments plus outputs less the number of constraints. The number of constraints imposed by the model equals the number of outputs and therefore the number of degrees of freedom cannot exceed the number of instruments. The simulation approach gives all available freedom to the instruments but, in optimisation terms, instruments and outputs are on an equal footing and should be treated as such.

The more general objective function introduces natural constraints on both instruments and targets and there are a number of efficient optimisation algorithms which automatically employ directional information at each step. Policy formulation via simulation (Steps 1–3) can be viewed as a single iteration within Step A in which directional information about desirable output values is lacking and is replaced by the skill of the modeller. However, no matter how long his experience or deep his grasp of economic theory, he will not always be able to correct his instruments in the right direction, given the size and complexity of most forecasting models. Thus his sequence of simulation exercises may actually fail to converge to the region of desirable policies by zigzagging around it. This is inefficient computationally and the end-result may well be an unnecessary compromise.

In general, the policy-maker would like to see a general nonlinear function representing his priorities. In most cases, however, such an objective function either cannot be specified or cannot be tailored to produce optimal values which the decision-maker regards as totally satisfactory.

A quadratic objective function is often assumed in order to overcome these difficulties. Other points in favour of such a choice are:

(1) It may be regarded as a second order Taylor series approximation to the 'true' objective function of the policy maker.
(2) It gives a simple measure of the weighted deviations of the actual variable values from their desired trajectories;
(3) The solution to the optimal control problem with a quadratic objective function is well defined and extensively researched.

Earlier attempts (e.g. Bray (1975)) to specify an objective function which truly represents the priorities of the policy-maker generate more questions than answers. Recent control exercises have thus been based on particular fixed objective functions without being able to close the gap between the optimal values so generated and those that the policy-maker would really like to see.

The requirement that the policy maker must attach 'cardinal' measures to marginal deviations of the paths of economic targets from desired paths is a major weakness of objective functions. There is, nevertheless, an alternative approach. Instead of requiring the policy-maker to answer a set of hypothetical questions about his preferences it is easier to confine questions to the actual paths of the target variables and instruments that are produced by some initial run. The problem then, if the policy maker has any reservations about the paths shown to him, is to translate these reservations back into the weights of the objective function. This is, in essence, what the technique described later in Chapter 6 of this volume does. The policy maker simply has to reveal to those carrying out the optimal control runs on the computer some set of preferred values for the targets and instruments for any time period. He can also express the strength of any preference by indicating that he would much prefer if one particular variable or set of variables were pushed well towards the more preferred value. The central point is that the policy-maker is never asked to put numerical values to his marginal preferences and this must be regarded as a major advance on what has, hitherto, been necessary if a sensible objective function was to be constructed.

5. FINAL OBSERVATIONS

The closed-loop solution in optimal control is designed for the specific purpose of attenuating the effects of random disturbances to the economy. Instead of proposed government expenditure being fixed at some assigned value, it could be made a function of the level of employment and the balance of trade in the future. Equally the target values for the money supply could be made a function of world inflation rates or the level of unemployment. In practice it would not be as crude as this since the optimal controls are calculated on the basis of the dynamics of the economic model, and so are likely to be a function of linear combinations of current and past values of both targets and instruments. As targets deviate from desired values, corrective action of an appropriate kind will be taken to bring the economy back towards its desired trajectory.

It is worth repeating, however, that this does not imply some immutable feedback law to which the policy-maker must adhere. The economy cannot be wired up like a thermostat on a central heating system and left at that. There will still remain considerable scope for discretionary action. In general, if priorities change with a change in government so will the feedback law. If the model is updated to correct for some structural change in the economy, the feedback law will change also. If the assumptions governing the paths for the exogenous variables are revised, the feedback law will have to be recalculated. The important point is the extent to which the feedback law is sensitive to shifts in any of these factors and this, of course, is ultimately an empirical matter.

It might be possible to design some broad control strategy, in keeping with the state of our knowledge, based on a small number of objectives and a limited number of instruments. This could be made sufficiently robust against the sorts of disturbances that have affected the UK economy since the war. It is worth noting, nevertheless, that in certain circumstances the feedback law might have to be recalculated. The quadrupling of crude oil prices in 1973 is a case in point. The size of this perturbation was probably greater than any experienced by the UK economy since the Korean War. If optimal control policies had been operating at the time, it is likely that they would not have been geared to cope with such a large shock, since the feedback law would have been calculated on the basis of variations in the balance of payments and import prices of the magnitude of those experienced over the previous 15 years. This of course does not rule out the possibility that the feedback law could be made robust enough to withstand a shock the size of the oil crisis.

A major stumbling-block to a wider acceptance of stochastic control methods is the difficulties of validation. When a control engineer has to convince industrialists that he can make a positive contribution to the solution of some of their problems he is able to do so, in the final analysis, by demonstrating that the performance of the boiler or the chemical process has been improved. This can be achieved in a relatively short space of time. In the management of the economy, however, the period necessary to demonstrate that stochastic control methods are beneficial may be prohibitively long.

It would be unreasonable to expect governments to present the economy as hostage to unproven methods, despite the attractions of the control framework on the theoretical level, and its success in industrial and military applications. Any scheme to introduce control methods into economic policy-making would need to proceed cautiously. The most that can reasonably be expected at this stage is that control methods will be allowed to prove themselves against present techniques of policy simulation and forecasting. Experience will suggest how best the new methods of control can be combined with those currently in use, and the best blend integrated into the existing framework of policy formulation. Unless, however, control methods are given an opportunity to demonstrate their effectiveness in the field no amount of prior theorising will be any use in the end.

PART II

6. INTRODUCTION

The technical sections that follow give a brief exposition of the role of state space concepts and optimisation theory in the modelling and control of economic systems. Much of modern control theory is couched in state space terms and this framework is particularly suitable when linear macroeconomic models are being used. On the other hand, attempts at directly controlling large nonlinear models mainly draw their inspiration from the broader field of constrained optimisation theory.

7. STATE SPACE FORMULATION

In dynamic system theory, models expressing the evolution of the variables of interest can be constructed in either discrete or continuous time. Discrete time models seem the more appropriate for economic systems: measurements are taken at discrete intervals (quarterly, semi-annually, etc.) and instruments are manipulated at discrete instants of time (remaining constant until the next manipulation).

Letting k denote the discrete time indexing set $k \in I = \{\ldots, -1, 0, 1, 2, \ldots\}$, such models may be expressed in the form

$$S: \underline{y}(k) = f(\underline{y}(k-1), \underline{y}(k-2), \ldots, \underline{y}(k-s), \underline{u}(k-1), \ldots, \underline{u}(k-r), k).$$

Here $\underline{y}(k)$ denotes an n-vector of endogenous variables (outputs), $\underline{u}(k)$ an m-vector of policy instruments (inputs) and f is an n-vector function, possibly nonlinear in all its arguments.

The explicit appearance of k in S above denotes a time-varying model. In general, such models involve time-dependent coefficients which can greatly complicate the computation of closed form solutions to S.

Conceptually, however, it should be noted that all the necessary theory for optimal control, prediction and filtering has been established for time dependent linear systems (Astrom, 1970).

The greater part of modern control theory deals with system models in *state space form.* An essential aspect of such forms in discrete time is that the system dynamics are described by a set of *first-order difference equations.* A simple example of a linear, time-invariant dynamic model in economics is the lagged multiplier-accelerator and this will suffice as an illustration:

$$Y(k) = C(k) + I(k) + U(k) \tag{7.1a}$$

$$C(k) = b\,Y(k-1) \tag{7.1b}$$

$$I(k) = a(C(k) - C(k-1)) \tag{7.1c}$$

where $Y(k)$, $C(k)$, $I(k)$, $U(k)$ are national income, consumption, investment and government expenditure respectively at quarter k.

From equations (7.1), $Y(k)$ satisfies

$$Y(k) = b(1+a)\,Y(k-1) - ab\,Y(k-2) + U(k) \tag{7.2}$$

a linear constant coefficient second order difference equation.

Alternatively, equations (7.1) can be written in the state space form:

$$\underline{x}(k+1) = F\underline{x}(k) + GU(k) \tag{7.3a}$$

$$Y(k) = H\underline{x}(k) + U(k) \tag{7.3b}$$

where the state vector $\underline{x}(k)$ at time k is given by

$$\underline{x}(k) = \begin{bmatrix} 1/b & -1/ab \\ 1/b & 0 \end{bmatrix} \begin{bmatrix} C(k) \\ I(k) \end{bmatrix} \tag{7.4}$$

and

$$F = \begin{bmatrix} 0 & 1 \\ -ab & b(1+a) \end{bmatrix}; \; G = \begin{bmatrix} 0 \\ 1 \end{bmatrix}$$
$$H = [-ab, \quad b(1+a)]. \tag{7.5}$$

The following points are important:

(i) The state space representation is not unique;

(ii) The state can be regarded as the 'memory' of the system; it stores a quantity of information which, together with the future input sequence, is sufficient to calculate the future output sequences of the system;

(iii) In contrast to the input-output representation (7.2), the state

space representation gives an internal description of the system, i.e. if the state $\underline{x}(k)$ is given, then $C(k)$, $I(k)$, $Y(k)$ can all be recovered;
(iv) If the state is eliminated from (7.3), then (7.2) results, i.e. both representations have the same input-output mapping;
(v) From (7.2) it follows that (7.1) is a stable system if the quadratic equation

$$\lambda^2 - b(1+a)\lambda + ab = 0 \tag{7.6}$$

has roots only inside the unit circle. This is equivalent to the system matrix F in (7.3) having eigenvalues only inside the unit circle;
(vi) The components of the state vector may not have any direct economic significance.

These points carry over to the dynamic analysis of more general systems.

8. SOME GENERALISATIONS

The state space representation (7.3) easily generalises to the multivariable case:

$$\underline{x}(k+1) = F\underline{x}(k) + G\underline{U}(k) \tag{8.1a}$$
$$\underline{Y}(k) = H\underline{x}(k) + D\underline{U}(k) \tag{8.1b}$$

where $\underline{Y}(k)$, $\underline{U}(k)$ are vectors.

The relevance of the state space matrices in (8.1) is clarified if the unit shift operator z is introduced, i.e. $z^{-j}\underline{U}(k) = \underline{U}(k-j)$. Then the system can be written in its final form representation

$$\underline{Y}(k) = T(z^{-1})U_k = \sum_{i=0}^{\infty} T_i\underline{U}(k-i) \tag{8.2}$$

where $T(z^{-1})$ is a matrix whose elements are rational polynomials in z^{-1} and

$$T(z^{-1}) = H(zI-F)^{-1}G + D \tag{8.3a}$$
$$= D + H\left[\sum_{i=1}^{\infty} F^{i-1}z^{-i}\right]G \tag{8.3b}$$

so that

$$T_o = D,\ T_1 = HG,\ T_2 = HFG, \ldots, T_k = HF^{k-1}G, \ldots \tag{8.4}$$

The final form matrices T_i in (8.4), expressed in terms of the state space matrices (F, G, H, D) are termed Markov matrices in the control literature. In the econometric literature, these matrices are referred to as matrix multipliers. Thus $T_0(=D)$ is designated the impact matrix multiplier, since its elements measure the immediate effect of a unit variation in any input on any output. Similarly, the matrix T_i is termed the interim, delay or dynamic matrix multiplier at lag i. Finally, assuming F is stable, $D+H[I-F]^{-1}G$ is the steady-state static matrix multiplier, representing the equilibrium effects on the outputs of a sustained unit change in the inputs.

Inspection of the T_i in (8.4) reveals that it is the state transition matrix F that determines their dynamic evolution. Thus the final form conceals dynamic information which is only made explicit in the state space form. The use of state space techniques for dynamic policy studies of multi-equation systems is therefore an explicit recognition of the proposition that the state space form best codifies dynamic structure.

Two further points are in order:

(i) Not every state space representation of the form (8.1) for a particular system has the same state dimension. Minimal representations (i.e. those of lowest state order) are assumed in what follows. These have certain desirable features but this important aspect of state space theory is not discussed here (see Caines and Wall, 1975; Preston and Wall, 1973a, 1973b).
(ii) State space models can be generalised readily to more complex situations (i.e. nonlinear, time-varying models, etc.) and are therefore of wide applicability.

9. CONTROL OF STATE SPACE MODELS

Consider the following problem:

Given the system

$$\begin{aligned} \underline{x}(k+1) &= F\underline{x}(k)+G\underline{U}(k);\ \underline{x}(1) \text{ initial state} \\ \underline{Y}(k) &= \underline{x}(k) \end{aligned} \tag{9.1}$$

choose a control sequence $\underline{U}(1)$, $\underline{U}(2)$, . . . so that $\lim_{k\to\infty} \underline{Y}(k) = \underline{0}$. Clearly, if F is asymptotically stable (i.e. its eigenvalues lie inside the unit circle) then the choice $\underline{U}(k)=0$, $k=1, 2, \ldots$ may be satisfactory in that the state will die away naturally. However, if F is unstable or has eigenvalues inside but close to the unit circle, then a better choice may take the form

$$\underline{U}(k) = K\underline{x}(k), \tag{9.2}$$

i.e. a *closed-loop linear state feedback control law.*

Substituting (9.2) in (9.1) yields

$$\underline{x}(k+1) = (F+GK)\underline{x}(k);\ \underline{x}(1) \text{ initial state} \tag{9.3}$$

and it is immediately clear that *the system dynamics have changed* in that the new system matrix is $F+GK$. If the representation (9.1) is minimal it can be shown that the matrix K can be chosen to arbitrarily allocate the eigenvalues of $F+GK$. In particular, if all the eigenvalues are placed at the origin, then $\underline{Y}(k) = \underline{0}$, $k \geqslant n$ where n is the state dimension ('deadbeat' controller) (Gopinath, 1971).

For each choice of K, (9.3) can be solved and (9.2) then yields

$$\underline{U}(k) = K(F+GK)^{k-1}\underline{x}(1) \tag{9.4}$$

i.e. an *open-loop control law.*

Note that, *in the presence of complete knowledge, the results obtained by using either (9.2) or (9.4) are completely identical.*

If uncertainty is present, however, the results obtained by using open-loop and closed-loop laws will not be the same. For example, if measurement error is present, then $\underline{x}(1)$ will only be an approximation to the true initial state. If F is unstable, then the error will grow without bound if the open-loop control law is used. The use of a suitable feedback law (9.2), however, ensures stability. In particular, a 'deadbeat' controller operates as before and brings the output to zero in n steps.

10. DETERMINISTIC OPTIMAL CONTROL

If the control law is chosen to achieve some desired output behaviour, this may lead to unacceptable inputs. In this case it is necessary to trade-off between conflicting requirements on the inputs and outputs of the system and this may be achieved via an objective function.

If the system is linear and the objective function quadratic, the basic deterministic problem is as follows:

Problem I: Given the system

$$\begin{aligned}\underline{x}(k+1) &= F(k)\underline{x}(k) + G(k)\underline{u}(k)\\ \underline{y}(k) &= H(k)\underline{x}(k) + D(k)\underline{u}(k)\quad k = 1, \ldots, N\end{aligned} \tag{10.1}$$

choose $\underline{u}(1), \ldots, \underline{u}(N)$ to minimise

$$J = \sum_{k=1}^{N} \{\delta\underline{y}(k)^T A(k)\delta\underline{y}(k) + 2\delta\underline{u}(k)^T C(k)\delta\underline{y}(k) + \delta\underline{u}(k)^T B(k)\delta\underline{u}(k)\}$$

where

$\delta\underline{u}(k)\underline{\Delta}\underline{u}(k) - \underline{u}(k)^d$, $\delta\underline{y}(k)\underline{\Delta}\underline{y}(k) - \underline{y}(k)^d$,
$\underline{u}(k)^d, \underline{y}(k)^d$ are desired values of $\underline{u}(k), \underline{y}(k)$,
$A(k)$ is positive semidefinite, $B(k)$ is positive definite
$(k = 1, \ldots, N)$.

The solution of this problem yields a feedback law of the form

$$\underline{u}(k) = L(k)\underline{x}(k) + \underline{h}(k) \tag{10.2}$$

where $L(k), \underline{h}(k)$ are the control gain and tracking gain respectively at time k. The sequence of gains can be derived via dynamic programming and depend on the system dynamics and the performance indices.

11. STOCHASTIC OPTIMAL CONTROL

Uncertainty can arise in a number of ways when modelling an economic system. The simplest method of incorporating stochastic effects is by introducing additive Gaussian noise. Thus (10.1) is replaced by

$$\begin{aligned} \underline{x}(k+1) &= F(k)\underline{x}(k) + G(k)\underline{u}(k) + \underline{w}(k) \\ \underline{y}(k) &= H(k)\underline{x}(k) + D(k)\underline{u}(k) + \underline{v}(k) \quad k = 1, \ldots, N \end{aligned} \tag{11.1}$$

where $\{w(k)\}_1^N$, $\{v(k)\}_1^N$ are independent Gaussian white noise sequences.

The LQG problem is then

Problem II: Given the system (11.1) choose $\underline{u}(1), \ldots, \underline{u}(N)$ to minimise $\bar{J}\underline{\Delta}E(J)$ where $E(.)$ denotes expectation.

The solution to this problem is given by

$$\underline{u}(k) = L(k)\hat{\underline{x}}(k/k-1) + \underline{h}(k) \tag{11.2}$$

where $L(k), \underline{h}(k)$ are identical to the gains in Problem I and $\hat{\underline{x}}(k/k-1)$ is the estimated value of the state based on input-output data up to time $k-1$.

The estimate $\hat{\underline{x}}(k/k-1)$ can be derived using the Kalman filter for (11.1) (provided the initial state and noise characteristics are known).

This is not pursued here but a useful review of Kalman filtering in econometrics has been written by Athans (1974).

The possibility of separating state estimation from control is a consequence of the *Separation Principle* and research is continuing in this area to determine the most general circumstances under which the principle remains valid. A particular case of the Separation Principle is *Certainty Equivalence* under which the control law is the same as in the deterministic case. This holds for the LQG problem as seen in (11.2).

In cases where the Separation Principle cannot be used, the computational load needed to solve the stochastic optimal control problem becomes very heavy.

12. A PARTICULAR STATE SPACE FORM

The work of the PREM group on the FIML estimation of linear models in rational structural form (Wall, 1975) led on to consideration of state space models of the form:

$$\begin{aligned} \underline{x}(k+1) &= F\underline{x}(k) + G_u\underline{u}(k) + G_e\underline{e}(k) + G_w\underline{w}(k) \\ \underline{y}(k) &= H\underline{x}(k) + D_u\underline{u}(k) + D_e\underline{e}(k) + D_w\underline{w}(k) \end{aligned} \tag{12.1}$$

where $\underline{e}(k)$, $k = 1, \ldots, N$ is a sequence of exogenous inputs (Preston and Wall, 1973a).

Although this is a stochastic model, its form implies *full state information*, thus avoiding the use of Kalman filtering techniques. This can be shown as follows.

Eliminating $\underline{w}(k)$ in (12.1) yields

$$\underline{x}(k+1) = A\underline{x}(k) + B\underline{u}(k) + C\underline{e}(k) + K\underline{y}(k) \tag{12.2}$$

where

$$\begin{aligned} A &= F - KH \quad ; K = G_w D_w^{-1} \\ B &= G_u - KD_u; \; C = G_e - KD_e \end{aligned}$$

and the matrix A is asymptotically stable (Preston & Wall, 1973a, p. 41). Therefore, if $\underline{x}(-d)$ is (incorrectly) set equal to zero for some positive d (usually at the historical origin), then the error that is committed will die away as (12.2) is run forward to the control origin. Thus the state $\underline{x}(1)$ at the control origin can be generated to an arbitrary accuracy. Once $\underline{x}(1)$ is known and $\underline{y}(1)$ is observed, then $\underline{x}(2)$ can be calculated and so on.

13. THE LQG CONTROL PROBLEM WITH FULL STATE INFORMATION

This section contains a brief summary of the solution to the optimal stochastic control problem for time-varying weightings when full state information is available. The detailed solution differs little from the constant weighting case (Wall and Westcott, 1975) and is included here for completeness.

The problem: Determine the vector of instruments $\{\underline{u}(k), 1 \leqslant k \leqslant N\}$ to minimise the cost function J given by

$$J = E\{\frac{1}{2}\sum_{k=1}^{N}[\delta\underline{y}(k)^T A(k)\delta\underline{y}(k) + 2\delta\underline{u}(k)^T C(k)\delta\underline{y}(k) + \delta\underline{u}(k)^T B(k)\delta\underline{u}(k)]\} \tag{13.1}$$

subject to

$$\underline{x}(k+1) = F\underline{x}(k) + G_u\underline{u}(k) + G_e\underline{e}(k) + G_w\underline{w}(k);\ \underline{x}(1) \text{ given} \tag{13.2}$$

$$\underline{y}(k) = H\underline{x}(k) + D_u\underline{u}(k) + D_e\underline{e}(k) + D_w\underline{w}(k);\ \underline{w}(k) \sim N(0,\Omega) \tag{13.3}$$

The matrices $B(k)$, $k = 1, \ldots, N$ are assumed to be symmetric and positive definite while the matrices $A(k)$, $k = 1, \ldots, N$ are assumed to be symmetric and positive semidefinite. The deviations $\delta\underline{y}(k)$, $\delta\underline{u}$(k) are defined by

$$\delta\underline{y}(k) = \underline{y}(k) - \underline{y}^d(k);\ \delta\underline{u}(k) = \underline{u}(k) - \underline{u}^d(k) \tag{13.4}$$

where the superfix d denotes the desired (or nominal) value.

Solution: The problem can be reduced to a standard state tracking problem if (13.3) is used to eliminate the endogenous variables from J. This leads to

$$\delta\underline{y}(k) = H\delta\underline{x}(k) + D_u\delta\underline{u}(k) \tag{13.5}$$

$$J = E\{\frac{1}{2}\sum_{k=1}^{N}[\delta\underline{x}(k)^T Q(k)\delta\underline{x}(k) + 2\delta\underline{u}(k)^T M(k)\delta\underline{y}(k) + \delta\underline{u}(k)^T R(k)\delta\underline{u}(k)]\} \tag{13.6}$$

where

$$Q(k) = H^T A(k) H \tag{13.7a}$$

$$M(k) = D_u^T A(k) H + C(k) H \tag{13.7b}$$

$$R(k) = B(k) + D_u^T A(k) D_u + C(k) D_u + D_u^T C(k)^T \tag{13.7c}$$

and

$$\underline{x}^d(k) = H^{\dagger}\{\underline{y}^d(k) - D_u \underline{u}^d(k) - D_e \underline{e}(k)\}$$
$$H^{\dagger} = \text{pseudoinverse of } H. \qquad (13.8)$$

The dynamic programming solution to the problem yields an optimal state feedback solution of the form

$$\underline{u}^*(k) = L(k)\underline{x}(k) + \underline{h}(k) \qquad (13.9)$$

and the optimal cost-to-go at time k is given by

$$V[\underline{x}(k), k] \,\underline{\Delta}\, \tfrac{1}{2}\underline{x}(k)^T P(k)\underline{x}(k) - \underline{a}(k)^T \underline{x}(k) + p(k) \qquad (13.10)$$

The arrays $L(k)$, $\underline{h}(k)$, $P(k)$, $\underline{a}(k)$, $p(k)$ are given by the following backwards recursion:

(1) Set $L(N) = -R(N)^{-1}M(N)$, $\underline{h}(N) = -L(N)\underline{x}^d(N) + \underline{u}^d(N)$
$P(N) = Q(N) + M(N)^T L(N)$, $\underline{a}(N) = P(N)\underline{x}^d(N)$,
$P(N) = \tfrac{1}{2}\underline{x}^d(N)^T P(N)\underline{x}^d(N)$

(2) For $k = N-1, N-2, \ldots\ldots, 2, 1$

$$\underline{\underline{R}}(k) \,\underline{\Delta}\, R(k) + G_u^T P(k+1)G_u, \quad \underline{\underline{M}}(k) \,\underline{\Delta}\, M(k) + G_u^T P(k+1)F$$

$$L(k) = -\underline{\underline{R}}(k)^{-1}\underline{\underline{M}}(k) \qquad (13.11)$$

$$\underline{h}(k) = -\underline{\underline{R}}(k)^{-1}[G_u^T P(k+1)G_e \underline{e}(k) - G_u^T \underline{a}(k+1) - M(k)\underline{x}^d(k) - R(k)\underline{u}^d(k)] \qquad (13.12)$$

$$P(k) = Q(k) + F^T P(k+1)F + L(k)^T \underline{\underline{M}}(k) \qquad (13.13)$$

$$\underline{a}(k) = F^T \underline{a}(k+1) - \underline{\underline{M}}(k)^T \underline{h}(k) - F^T P(k+1)G_e \underline{e}(k) + Q(k)\underline{x}^d(k) + M(k)^T \underline{u}^d k) \qquad (13.14)$$

$$p(k) = p(k+1) + \tfrac{1}{2}tr[P(k+1)G_w \Omega G_w^T] + \tfrac{1}{2}\underline{x}^d(k)^T Q(k)\underline{x}^d(k) + \underline{u}^d(k)(k)M(k)\underline{x}^d(k) + \tfrac{1}{2}\underline{u}^d(k)^T R(k)\underline{u}^d(k) - \underline{a}(k+1)^T G_e \underline{e}(k) - \tfrac{1}{2}\underline{h}(k)^T \underline{\underline{R}}(k)\underline{h}(k) + \tfrac{1}{2}\underline{e}(k)^T G_e^T P(k+1)G_e \underline{e}(k) \qquad (13.15)$$

From (13.10), the cost incurred over the optimal trajectory is

$$J^* = \tfrac{1}{2}\underline{x}(1)^T P(1)\underline{x}(1) - \underline{a}(1)^T x(1) + P(1) \qquad (13.16)$$

Note that the above solution is valid for the case when the model (13.2), (13.3) is time-varying; in this case the system matrices appearing on the right hand sides of equations (13.11)–(13.15) are replaced by their values at time k.

The model structure (13.2) and (13.3) thus implies both a simple method of recovering the state via (12.2) and a linear optimal feedback law (13.9) for which the gains can be computed and stored 'off-line', i.e. once and for all at the control origin. This leads to a relatively simple scheme for controlling a large nonlinear model using a small linear model which is discussed in Chapter 10.

14. GENERAL APPROACH TO THE DERIVATION OF NEWTON-TYPE ALGORITHMS FOR THE OPTIMAL CONTROL OF ECONOMETRIC MODELS

The task of controlling large and nonlinear econometric models for the purposes of economic policy evaluation has led to the design of a number of simple optimisation algorithms. The open-loop control algorithms of Holbrook (1974) and Fair (1974) and the closed loop approach of Chow (1975) are well-known examples. Apart from Fair (1974), all other attempts have been confined to quadratic objective functions.

Preston *et al.* (1976) have devised a simple algorithm for controlling the Wharton Long Term Annual and Industry Forecasting Model. This algorithm takes the quadratic nature of the objective function fully into account in its iterative procedure.

A quadratic objective function can be considered equivalent to the second order Taylor series expansion about the unconstrained optimum of the true nonlinear objective function. The constant term of this expansion can be ignored without affecting the value of the optimal trajectory. Consider, therefore, the quadratic objective function

$$J(\underline{Y}, \underline{U}) = \tfrac{1}{2}(\underline{Y} - \underline{Y}^d)^T Q_y (\underline{Y} - \underline{Y}^d) + \tfrac{1}{2}(\underline{U} - \underline{U}^d)^T Q_u (\underline{U} - \underline{U}^d) \quad (14.1)$$

where

$$\underline{Y}^T = (\underline{y}^T(1), \underline{y}^T(2), \ldots, \underline{y}^T(k), \ldots, \underline{y}^T(\mathrm{N})) \quad (14.2)$$

is the mN vector of endogenous variables throughout the period of optimisation [1, N], with $\underline{y}(k)$ denoting the m-vector of endogenous variables at time k, $1 \leqslant k \leqslant \mathrm{N}$, and

$$\underline{U}^T = (\underline{u}^T(1), \underline{u}^T(2), \ldots, \underline{u}^T(k), \ldots, \underline{u}^T(\mathrm{N})) \quad (14.3)$$

is the nN-vector of policy instruments (controls) written in the same form as (14.2) with $\underline{u}(k)$ as an n-vector. The superscript d denotes the desired values, the symmetric matrix Q_y is assumed to be positive semi-definite and the symmetric matrix Q_u is assumed to be strictly positive definite. The elements of these matrices penalise the departure of a variable from its desired trajectory. The off-diagonal elements indicate the measure of importance attached to the deviation in one variable versus the deviation in another.

The optimisation problem may thus be formulated as the choice of the endogenous variables $\underline{Y}$, and the policy instruments $\underline{U}$ such that the objective function (14.1) is minimised subject to the equations of the nonlinear econometric model throughout the period [1, N]. If these equations are written in compact form as $\underline{F}(\underline{Y}, \underline{U}) = \underline{0}$, the corresponding problem becomes

$$\min\{J(\underline{Y}, \underline{U}) | \underline{F}(\underline{Y}\ \underline{U}) = \underline{0}\}. \tag{14.4}$$

It is assumed that the values of all the exogenous variables which are not policy instruments and hence not subject to optimisation are substituted in $\underline{F}(\underline{Y}, \underline{U})$ before solving (14.4). The desired trajectories $\underline{Y}^d$, $\underline{U}^d$ are also the minima of the Taylor series expansion (14.1). As this expansion is about the unconstrained optimum of the true nonlinear objective function, $\underline{Y}^d$ and $\underline{U}^d$ are exactly the values of $\underline{Y}$, $\underline{U}$ which minimise the true welfare function. The vectors $\underline{Y}^d$, $\underline{U}^d$ may, therefore, be thought of as desired but possibly infeasible values considering the restriction imposed by the model. The constrained optimisation problem (14.4) can thus be interpreted as getting as near as possible to these desired values. Clearly, if the desired trajectories are feasible and hence satisfy $\underline{F}(\underline{Y}^d, \underline{U}^d) = \underline{0}$ then they are also the solution of (14.4).

Having formulated what is basically a dynamic problem in terms of static optimisation, the solution will be derived in the Euclidean spaces $E^{m\mathrm{N}}$ and $E^{n\mathrm{N}}$ for $\underline{Y}$ and $\underline{U}$ respectively. Such transcriptions of dynamic optimisation problems are extensively discussed by Canon, Cullum, Polak (1970) and Polak (1971).

Assumption (14.5)

Given $\underline{U} \in E^{n\mathrm{N}}$, the set of equations $\underline{F}(\underline{Y}, \underline{U}) = \underline{0}$ may be solved for $Y \in E^{m\mathrm{N}}$. Thus, there exists a mapping $g: E^{n\mathrm{N}} \to E^{m\mathrm{N}}$ such that $\underline{Y} = \underline{g}(\underline{U})$.

Remark (14.6)

For every econometric model, the solution program computes the endogenous variables given the exogenous values. Hence the mapping $\underline{g}$ is provided by the solution program of an econometric model.

Assumption (14.5) may be used to eliminate $\underline{Y}$ from (14.1), thus

$$\begin{aligned} G(\underline{U}) &= J(\underline{g}(\underline{U}), \underline{U}) \\ &= \tfrac{1}{2}(\underline{g}(\underline{U}) - \underline{Y}^d)^T Q_y (\underline{g}(\underline{U}) - \underline{Y}^d) + \tfrac{1}{2}(\underline{U} - \underline{U}^d)^T Q_u (\underline{U} - \underline{U}^d) \end{aligned} \tag{14.7}$$

where the first term on the right-hand side of (14.7) may be seen as a weighted least squares term. The original constrained problem (14.4) may now be formulated as an equivalent unconstrained optimisation problem in the policy instruments only, i. e.

$$\min\{G(\underline{U})|\underline{U}\in E^{nN}\} \tag{14.8}$$

The minimum in (14.8) can be found iteratively by successively generating suitable descent directions.

If G is assumed approximately quadratic, then

$$\nabla G(\underline{U}_k)-\nabla G(\underline{U}_{k+1})=H_{\underline{U}_k}(\underline{U}_k-\underline{U}_{k+1}) \tag{14.9}$$

where $\underline{U}_k, \underline{U}_{k+1}\in E^{nN}$ are two policy instrument vectors, $\nabla G(\underline{U}_k)$ is the gradient of (2.7) evaluated at $\underline{U}_k$ and $H_{\underline{U}_k}$ is the Hessian of (14.7) evaluated at $\underline{U}_k$. The subscripts k and $k+1$ are at this stage used to indicate two successive and distinct values of the policy instrument vector. The descent direction known as the Newton-type direction is obtained by assuming that $\underline{U}_{k+1}$ is the unconstrained optimum of $G(\underline{U})$ (i.e. the solution of (14.8)) for which the necessary condition is

$$\nabla G(\underline{U}_{k+1})=\underline{0} \tag{14.10}$$

which may be substituted into (14.9) to obtain

$$\underline{U}_{k+1}=\underline{U}_k-H_{\underline{U}_k}^{-1}\nabla G(\underline{U}_k) \tag{14.11}$$

assuming the Hessian to be positive definite. In general, $G(\underline{U})$ is nonquadratic in $\underline{U}$ and (14.11) is modified with the scalar

$$\underline{U}_{k+1}=\underline{U}_k-\alpha H_{\underline{U}_k}^{-1}\nabla G(\underline{U}_k) \tag{14.12}$$

such that $G(\underline{U}_{k+1})\leqslant G(\underline{U}_k)$ is maintained. Note that the direction $\underline{U}_{k+1}-\underline{U}_k$ is a descent direction at $\underline{U}$, i.e.

$$\langle\nabla G(\underline{U}_k), \underline{U}_{k+1}-\underline{U}_k\rangle=-\alpha\langle\nabla G(\underline{U}_k), H_{\underline{U}_k}^{-1}\nabla G(\underline{U}_k)\rangle\leqslant 0 \tag{14.13}$$

Hence for some $\alpha>0$ a decrease in cost is assured.

Given (14.7), the gradient and Hessian of $G(\underline{U})$ at $\underline{U}_k$ may be written explicitly. Thus

$$\nabla G(\underline{U}_k)=\mathcal{N}_k^T Q_y(\underline{g}(\underline{U}_k)-\underline{Y}^d)+Q_u(\underline{U}_k-\underline{U}^d) \tag{14.14}$$

where

$$\mathcal{N}_k \triangleq \left.\frac{\delta \underline{g}}{\delta \underline{U}}\right|_{\underline{U}=\underline{U}_k}=\left.\frac{\delta \underline{Y}}{\delta \underline{U}}\right|_{\underline{U}=\underline{U}_k} \text{ with } \frac{\delta Y_i}{\delta U_j}=\left(\frac{\delta \underline{Y}}{\delta \underline{U}}\right)_{ij} \tag{14.15}$$

and

$$H_{\underline{U}_k} = \mathcal{N}_k^T Q_y \mathcal{N}_k + B_k + Q_u \tag{14.16a}$$

$$B_k = \sum_{i=1}^{mN} \nabla^2 g_i(\underline{U}_k) Q_y (g_i(\underline{U}_k) - Y_i^d) \tag{14.16b}$$

Y_i^d and g_i are the ith elements of their corresponding vectors, $\nabla^2 g_i(\underline{U}_k)$ is the Hessian of g_i with respect to $\underline{U}$, evaluated at $\underline{U}_k$.

A Gauss-Newton type method may be obtained by neglecting the second derivative term B in (14.16). It may be argued that econometric models are not very nonlinear so that $||B||$ will be small compared to $||\mathcal{N}_k^T Q_y \mathcal{N}_k + Q_u||$, and thus ignoring B is justified.

Remark (14.17)

In the original Gauss-Newton algorithm (see Ortega and Rheinboldt (1970)) Q_u is absent and therefore, to avoid singularity problems, B cannot be ignored (see Gill and Murray (1976)). It should be noted that if the matrix B is omitted in (14.16) then $H_{\underline{U}_k}$ may be written as

$$H^o_{\underline{U}_k} = \mathcal{N}_k^T Q_y \mathcal{N}_k + Q_u \tag{14.18}$$

evaluated at any $\underline{U}_k \in E^{nN}$, $k = 0, 1, \ldots$ and $H^o_{\underline{U}_k}$ always remains positive definite since $Q_y \geqslant 0$ and $Q_u > 0$. Hence the direction

$$-H^{o^{-1}}_{\underline{U}_k} \nabla G(\underline{U}_k) \tag{14.19}$$

is a descent direction at $\underline{U}_k$ since

$$-< \nabla G(\underline{U}_k), H^{o^{-1}}_{\underline{U}_k} \nabla G(\underline{U}_k) > \leqslant 0. \tag{14.20}$$

Thus an algorithm may be constructed by using

$$\underline{U}_{k+1} = \underline{U}_k - \alpha H^{o^{-1}}_{\underline{U}_k} \nabla G(\underline{U}_k) \tag{14.21}$$

iteratively while choosing the scalar α to ensure that $G(\underline{U}_{k+1}) \leqslant G(\underline{U}_k)$. Using (14.14) and (14.18) an explicit formula for (14.21) may be obtained,

$$\underline{U}_{k+1} = \underline{U}_k - (\mathcal{N}_k^T Q_y \mathcal{N}_k + Q_u)^{-1} (\mathcal{N}_k^T Q_y (\underline{Y}_k - \underline{Y}^d) + Q_u (\underline{U}_k - \underline{U}^d)). \tag{14.22}$$

Note that the matrix $(\mathcal{N}^T Q_y \mathcal{N} + Q_u)$ is $n\mathrm{N} \times n\mathrm{N}$ and is easily invertible since the number of policy instruments is usually not very high.

15. SOME EXISTING ALGORITHMS

The algorithm proposed and implemented by Preston *et al.* (1976) can be easily derived from the analysis of the previous section. When $\underline{U}_k$ is the required constrained optimum of (14.4), the gradient (14.14) is equal to zero. As stated earlier, this is a necessary condition for the unconstrained optimum of $G(\underline{U})$. Hence, using (14.14)

$$\mathcal{N}_k^T Q_y(\underline{g}(\underline{U}_k) - \underline{Y}^d) + Q_u(\underline{U}_k - \underline{U}^d) = \underline{0}$$

and

$$\underline{U}_k = \underline{U}^d - Q_u^{-1} \mathcal{N}_k^T Q_y(g(\underline{U}_k) - \underline{Y}^d) \tag{15.1}$$

Starting with an initial value for $\underline{U}_k$ and updating via (15.1) leads to the basic idea underlying the iterative procedure used by Preston *et al.* (1976), thus

$$\underline{U}_{k+1} = \underline{U}^d - Q_v^{-1} \mathcal{N}_k^T Q_y(g(\underline{U}_k) - \underline{Y}^d) \tag{15.2}$$

The algorithm used by Preston *et al.* (1976) constrains Q_y and Q_u to diagonal matrices. Such a restriction has not been imposed here.

The iterative procedure (14.11) or its simplified version (14.21) with $\alpha \equiv 1$ is the basis of two other existing algorithms for the optimal control of econometric models. The algorithms developed by Holbrook (1974) and Chow (1975) both use a locally linearised model and a quadratic objective function. Chow, however, employs a dynamic programming procedure at the kth linearisation to derive $\underline{U}_{k+1}$. In the absence of stochastic disturbances in the model, Chow's dynamic optimisation procedure can clearly be expressed in terms of static optimisation (see, e.g., Polak (1971), p. 3). Thus, this algorithm is conceptually equivalent to (14.21) with $\alpha \equiv 1$.

REFERENCES

Åström, K. J., *Introduction to Stochastic Control Theory* (Academic Press, 1970).

Athans, M., 'The Role and Use of the Stochastic Linear–Quadratic–Gaussian Problem in Control System Design', *IEEE Trans.*, A.C. 16, 6, (Dec 1971).

Athans, M., 'The Importance of Kalman Filtering Methods in Economic Systems', *Annals of Social and Economic Measurement*, vol. 3, no. 1 (Jan 1974).

Athans, M., *et al.*, 'Sequential Open-Loop Optimal Control of a Nonlinear Macroeconomic Model', in *Frontiers of Quantitative*

Economics, ed. M. D. Intriligator (North-Holland, Amsterdam, 1976).

Bray, J., 'Optimal Control of a Noisy Economy with the U.K. as an Example', *J. R. Statist. Soc.*, A., 1975.

Caines, P. E., and Wall, K. D., 'Theoretical Foundations for Methods of Estimation and Control in Economic Policy Optimization', in *Modelling the Economy*, ed. G. A. Renton (SSRC, 1975).

Canon, M. *et al.*, *Theory of Optimal Control and Mathematical Programming* (McGraw-Hill, New York, 1970).

Chow, G. C. *Analysis and Control of Dynamic Economic Systems* (John Wiley and Sons, New York, 1975).

Fair, R., 'On the Solution of Optimal Control Problems as Maximisation Problems', *AESM*, vol. 3, 1 (Jan 1974) pp. 135–54.

Gill, P. E., and Murray, W., 'Algorithms for the solution of the nonlinear least-squares problem', National Physical Laboratory, Report NAC 71, (Teddington, England, 1976).

Gopinath, B., 'On the Control of Multiple Input-Output Systems', *The Bell System Telephone Journal*, vol. 50, no. 3 (Mar 1971).

Holbrook, R. S., 'A practical Method for Controlling a Large Nonlinear Stochastic System', *AESM*, vol. 3, 1 (Jan 1974) pp. 155–76.

Ortega, J. M., and Rheinboldt, W. C., *Iterative Solution of Nonlinear Equations in Several Variables* (Academic Press, New York and London, 1970).

Polak, E., *Computational Methods in Optimisation* (Academic Press, New York and London, 1971).

Preston, A. J., and Wall, K. D., 'An Extended Identification Problem for State Space Representations of Econometric Models', PREM Discussion Paper 6 (1973a).

Preston, A. J., and Wall, K. D., 'Some Aspects of the Use of State Space Models in Econometrics', PREM Discussion Paper 5 (1973b).

Preston, R. S., *et al.*, 'Control Theory Simulations Using the Wharton Long Term Annual and Industry Forecasting Model', *Wharton E.F.A.* (1976).

Wall, K. D., 'FIML Estimation of Rationa. Distributed Lag Structural Form Models', PREM Discussion Paper 11 (1975).

Wall, K. D., and Westcott, J. H., 'Policy Optimization Studies with a Simple Control Model of the U.K. Economy', PREM Discussion Paper 9 (Mar 1975).

PART II
ASPECTS OF THEORY AND POLICY

2 Effective Use of Econometric Models in Macroeconomic Policy Formulation

Gregory C. Chow[1]

At the beginning of each year, the *Economic Report of the President* of the United States makes projections of GNP in nominal and real terms for the coming year, the unemployment rate and the inflation rate, and states the major fiscal and monetary policies required to achieve these target rates. For example, the *Report* of January 1976 estimates real GNP to be 6 to 6.5 per cent higher in 1976 than in 1975 (p. 19), the unemployment rate to fall by almost a full percentage point and the inflation rate measured by the rise in the GNP deflator to be about 6 per cent (p. 24). The associated fiscal policies include a proposed Federal outlay in fiscal 1977 of \$394 billion, a cut in taxes beginning in July 1976 of about \$28 billion relative to what they would be under the 1974 law (p. 22). The rate of growth in the money supply M_1, as announced by the Federal Reserve, ranges between $5\frac{1}{2}$ and $7\frac{1}{2}$ per cent, but the *Report* asserts that maintaining a rate of money growth at the upper limit of this range would hinder the progress toward lower inflation rates (pp. 21–2). Assuming that econometric models are being used for policy analysis, this paper presents a systematic approach to apply some recently developed techniques of stochastic control to improve the formulation of macroeconomic policies and the accompanying economic projections.

The analysis starts with the tentative paths for the policy variables[2] which result from the existing procedure without the benefits of stochastic control methods. Although we assume that an econometric model is used, its inaccuracies will be duly considered. The recommended procedure consists of twelve steps.

1. Insert the tentative paths of the policy variables and the best available estimates of the exogenous variables not subject to control into the econometric model to obtain projections of the key economic variables for eight quarters. This step is already being performed in Great Britain, since the Treasury is required to maintain an econometric model in the public domain and to make and publish projections from the model, given the current policy proposals.

2. Modify the econometric model, the estimates of the uncontrollable exogenous variables, and/or the economic projections if the projections from step 1 differ from those obtained from whatever existing procedure used in the formulation of macroeconomic policies. When making forecasts, econometric forecasters in the United States adjust the constants in their model utilising observations of the equation residuals in recent quarters and other information. Others might insist on forecasting without adjustment of the model, in which case only the estimates of the uncontrollable exogenous variables and the final economic projections can be changed. Whatever adjustments of model parameters and economic projections are made, the essence of step 2 is to arrive at a set of forecasts of the important endogenous variables y_t^o, a set of estimates for the future uncontrollable variables z_t^o and an econometric model which are consistent with one another, given the tentative paths for the control variables x_t^o. Thus these variables satisfy each of the p simultaneous structural equations in the model

$$y_{it}^o = \Phi_i(y_t^o, y_{t-1}^o, x_t^o, z_t^o) + \varepsilon_{it} \, (i = 1, \ldots, p) \tag{1}$$

if the random residual ε_{it} is set equal to zero.

3. Set target values for the future unemployment rate, inflation rate, real GNP, measures of balance of payments and possibly other important economic variables which are somewhat more desirable than the values given by y_t^o in step 2. The motivation here is to find out whether the tentative path x_t^o for the policy variables can be improved upon by performing optimal control calculations. To do so, we choose a quadratic loss function and use the above target values as elements in the vector a_t:

$$W = \sum_{t=1}^{T} (y_t - a_t)' K_t (y_t - a_t) \tag{2}$$

where K_t is diagonal matrix giving weights according to the relative importance of the target variables, and the planning horizon T can be set equal to about 20 quarters.[3]

4. Linearise the equations (1) about the tentative paths y_t^o and x_t^o, given

z_t^o, obtaining a linear model of time-dependent coefficients, and compute the optimal feedback control equations

$$x_t = G_t y_{t-1} + g_t \tag{3}$$

which minimise the expectation of the loss function (2) subject to the constraint of the linear stochastic model. A computer program is available for this purpose, as described in Chow (1976b).

Briefly, the computer program applies the Gauss-Seidel iterative method to solve the possible nonlinear econometric model for y_t^o, given z_t^o and x_t^o, as required in step 1 above. It automatically linearises the nonlinear structural equations (1) which are input to the program in Fortran code, and solves the resulting linear structural equations to obtain a set of linear reduced-form equations

$$y_t = A_t y_{t-1} + C_t x_t + b_t + u_t \tag{4}$$

where the intercepts b_t incorporate the effects of z_t and the vectors of random residuals u_t are related to the residuals ϵ_{it} of (1) in a well-known manner. Then the coefficients G_t and g_t of the optimal feedback control equations (3) are computed. In the above notation, the vector y_t includes variables introduced to eliminate endogenous variables lagged more than one period and includes x_t as a subvector so that the loss function (2) has only y_t as argument.

By the use of this computer program after a set of optimal feedback control equations is obtained, a new set of y_t^o will be calculated to correspond to the new policies, and the nonlinear model will be linearised around the new tentative paths for y_t^o and x_t^o, yielding a new set of linear reduced-form equations (4). Another set of optimal feedback control equations are obtained, and the computations are repeated until the process converges. Our experience with several US models, including the Klein-Goldberger model, the St. Louis model and the University of Michigan Quarterly Econometric Model, is that it takes about three rounds of linearisations to converge. The Michigan model contains 61 endogenous variables from the original simultaneous equations plus 71 more new endogenous variables to convert the system into first-order, plus three variables $y_{133,t} = x_{1,t}, \ldots, y_{135,t} = x_{3,t}$ which are equal to the three control variables selected for our experiments, giving a vector of 135 elements for y_t in (1). To compute the optimal solution in one round of linearisations, using the Michigan model for a 17-period control problem with three control variables, costs about \$20.00 at the Princeton University Computer Center equipped with an IBM 360–91 Computer. The cost is expected to be about seven times (somewhat less than 2^3) if the size of the model doubles. If the number of planning periods changes, the cost will change linearly because the program takes advantage of the time structure of the problem and computes the feedback coefficients G_t and g_t period by period. Control algorithms which treat a minimisation problem with respect to the total number of variables (equal to the number of control variables times the number of periods) without regard to the time structure of the optimisation problem will become much more than twice as expensive when the number of variables doubles. Our program has the additional property that its cost will hardly increase at all when the number of control variables increases.

5. Change systematically the weighting matrix K_t and the targets a_t in the loss function (2) and reoptimise in order to trace out the best combinations of the future inflation rates and unemployment rates attainable given the econometric model. The procedure is described in detail in Chow and Megdal (1976). Essentially, if the weight $k_{11,t}$ corresponding to the unemployment rate y_{1t} is very large as compared with the weight $k_{22,t}$ for the inflation rate $y_{2t,}$ and if the target a_{1t} for y_{1t} is set at 5 per cent and the target for the inflation rate is set low enough, the solution will give the lowest inflation rate attainable for a 5 per cent unemployment. By varying a_{1t} from 4 to 8 per cent, one can compute the optimum solutions to find out the best inflation rates corresponding to these various unemployment rates and the associated policies required to achieve them. Since we are dealing with T periods, it may be useful to plot the mean unemployment and inflation rates over these periods, or to plot the root mean squared deviations of these rates form their targets.

6. Present the results of step 5 to the policy-makers who will then make a choice among the best feasible combinations of unemployment and inflation. It is quite likely that the unemployment and inflation rates from the tentative solution in step 1 are dominated by the solutions obtained in step 5. If the solution for the unemployment rate is around 6 per cent in step 1, say, the solution in step 5, using $a_{1t} = 6$ per cent, guarantees that the resulting inflation rates are the lowest possible as a consequence of optimisation. The choice made here and the corresponding optimal policy will constitute a set of intermediate solution paths for y_t and x_t for further analysis and improvement.

7. If the solution paths for the policy variables in step 6 drift very far away from the paths in step 1 or show severe fluctuations, impose penalties in the loss function for them and reoptimise. The weights in the K_t matrix may be assigned to the levels of the policy variables which are

given certain reasonable target paths. Or the quarter-to-quarter changes in some policy variables can be dampened by introducing the first differences as new variables which are then given appropriate weights in the K_t matrix and steered toward the target zero. Perhaps trials and errors are required in this step to obtain reasonable solution paths for the control variables.

8. Examine the reasonableness of the new solutions for y_t and x_t in step 7 using any outside information available. Adjust the econometric model and reoptimise if necessary. The need to adjust the econometric model and/or the estimates of the uncontrollable exogenous variables may arise at this stage because the new solutions in step 7 may be quite far from the solutions in step 1, affecting the accuracy of the econometric model as an approximation of reality and even conceivably affecting the values of some variables which have been treated as exogenous but may indeed react to sizeable changes in policies. Reoptimise after the model is adjusted.

9. If a second reasonable econometric model is available, it would be useful to apply the policy paths in step 8 to it and compare its projections of unemployment, inflation and real GNP with those obtained from the first model in step 8. If the two sets of projections are similar, or if the second set is as satisfactory as the first set from the original model (so that there is no risk of very bad performance if the alternative model is true), conclude the search for optimal policies and go to step 11. Otherwise, go to step 10.

10. Examine the consequences of at least three policies, (*a*) the optimal policy based on the first model as obtained in step 9, (*b*) the optimal policy based on the second model using the same loss function to be similarly computed, and (*c*) the originally proposed policy used in step 1, under the alternative assumptions that one of the two models is correct. Here a 3 by 2 payoff matrix can be utilised, with three policies combined with two possible states of the world or models. By applying the three policies to the two models, we can compute the total expected losses for eight quarters, say, to be entered in the above payoff matrix. If policy (*a*) or (*b*) dominates policy (*c*), as shown by the first or second row of the matrix having smaller losses than the third row, we have found an improvement over the policy originally proposed. If neither policy (*a*) nor (*b*) dominates (*c*), the payoff matrix will still serve as a useful tool of analysis. If one takes the Bayesian approach, he assigns probabilities to the two models and chooses that policy which minimises the expected loss obtained by weighing the losses from the policy by the probabilities. If one is conservative, he may choose the minimax strategy. An

illustrative analysis using such a payoff matrix can be found in Chow (1976c).

What if the two models disagree, as shown by large expected losses in the 1–2 and 2–1 entries in the above matrix, and one is unwilling to take the Bayesian approach to resolve the conflict? A further analysis can be performed. It is based on the idea that policies are made sequentially period by period, and that the policy-maker does not have to follow the policy recommendations computed from one model for many future periods after he decides to follow it for a quarter or two. The analysis described in the last paragraph ignores the possibility of shifting and revising models as it examines the expected total loss for many periods when the policy recommendations from one model are followed throughout. The disagreements between the policy recommendations from two different models would be reduced and the difficulties in choosing between conflicting policies would diminish if this possibility is taken into account. The first-period policies from the two models may not differ by very much even if following the recommendations from the two models for many periods would lead to very different consequences. Furthermore, assuming that the first-period policies based on the two models differ greatly, and that their multi-period expected losses also differ, the policy-maker would still not face a serious dilemma if he knows that following the policies from model 1 for one or two quarters and shifting to the policies from model 2 afterwards will be nearly as good as following the policies from model 2 for all periods when model 2 happens to be the true model.

In essence the i-j entry of payoff matrix in this analysis should show the total expected loss for many periods if the policy recommendation from model i is followed *only for period* 1 but the policies from model j will be followed afterwards.This construction is based on the notion that the decision for the first quarter, even if it is mistaken, can be corrected in the following quarters. Therefore, the damage done in this quarter is measured by the difference between the multiperiod losses incurred when (1) following the wrong policy of model i for one quarter but the correct policies of the right model j afterwards; and (2) following the policies of the correct model j all through, the latter being given by the j-j entry of the payoff matrix. Such a matrix is quite easy to compute if the optimal stochastic control algorithm described in step 4 is used. This algorithm is derived from the method of dynamic programming (Chow, 1975, Chapter 8) by which one reduces successively the problem of minimising the expected loss for T periods to the problem of minimising the expected loss for one period, starting with the problem for period T, and then the problem for the last two periods, etc., until the problem for all T periods is solved. The final problem amounts to minimising the expectation of a quadratic function $y_1'Hy_1 - 2y_1'h + c$ of only the variables y_1 in period one with respect to the first-period policy x_1, it being understood that, whatever the outcome y_1 for period one turns out to be, the future policies $x_2, \ldots, x_T$ shall be optimally chosen. See Chow (1975, pp. 178–9). Using the right-hand side of equation (4) to substitute for y_1 in the above quadratic loss function and taking its expectation, we find the total expected loss for T periods to be a quadratic function of x_1, say $x_1' Q_j x_1 - 2x_1' q_j + d_j$, where the subscript j indicates that the optimal control calculations are performed using model j. This function gives the expected T-period loss if x_1 is applied in the first period *and* x_2 to x_T shall be optimally chosen according to model j, under the assumption that model j is true. If we minimise this function with respect to x_1, we obtain the optimal first-period policy according to model j. If we apply the three different first-period policies x_1 used in the construction of the 3 by 2 payoff matrix to evaluate this function, we will obtain the entries for the jth column of the required payoff matrix.

The purpose of step 10 is to arrive at a final policy recommendation for the current quarter. Even the payoff matrix constructed in the last paragraph may show seriously conflicting first-period recommendations from the two alternative models, but a decision has to be reached by the Bayesian, Minimax or some other criterion. It is better to know the various risks involved under the alternative states of the world when making a decision than not to know them at all. When faced with conflicting recommendations, one may attempt to find a robust policy which would work reasonably well under the alternative models. This is a subject requiring further research. One approach is to modify the optimal policies by allowing for the uncertainty in the estimated parameters of the econometric models used, as described in Chow (1976a).

11. Calculate the mean paths and the covariance matrix of the major economic variables using a reasonable model and the optimal feedback control policy chosen above. The decision-makers should be informed of the likely consequences in the future when the recommended policy is applied. Using equations (3) and (4) obtained in step 4 above, we obtain a linear approximation of the dynamic stochastic system under control

$$\begin{aligned} y_t &= (A_t + C_t G_t)\, Y_{t-1} + (b_t + C_t g_t) + u_t \\ &= R_t\, y_{t-1} + r_t + u_t \qquad (5) \end{aligned}$$

The mean path of this system is given by

$$\bar{y}_t = R_t y_{t-1} + r_t \tag{6}$$

Using $y_t^* = y_t - \bar{y}_t = R_t y_{t-1}^* + u_t$, we can compute the covariance matrix by

$$Ey_t^* y_t^{*\prime} = R_t (Ey_{t-1}^* y_{t-1}^{*\prime}) R_t' + Eu_t u_t'$$

where the covariance matrix $Eu_t u_t'$ of the reduced form residuals are calculated from the estimated covariance matrix of the residual ε_t in the structural equations (1).

12. If the above steps are followed each quarter, the econometric models used will be revised and improved, and more weights will eventually be given to the recommendations from the models that have shown a better track record.

Why should the procedure outline above be adopted? It will make explicit the underlying rationale in the making of macroeconomic policies. If such an approach is not used, one would wonder on what basis government macroeconomic decisions are reached, what dynamic relationships among the important economic variables are assumed in policy-making, and what objectives the government is trying to achieve. Once these questions are answered explicitly and quantitatively, the logical approach is to write down the dynamic economic equations and the objective function, and to find the policies that would best achieve the objectives. This is precisely our recommendation. We have simply filled in the details in implementing such an approach by bringing the available econometric knowledge to bear and by designing a computationally efficient procedure to find optimal policies which are to be made sequentially and to ascertain the economic consequences of such policies.

The reader will have recognised that, although we suggest the use of stochastic control techniques for policy analysis, we are far from advocating the automatic use of these techniques without the intervention of human judgement and political considerations. Needless to say, poor policy recommendations are likely to follow from poor econometric models, no matter whether optimisation techniques are used or not. In reality, technical economic advice may play only a limited role in the formulation of economic policies. Whatever its limited role, the current practice has already incorporated the use of econometric models to simulate the likely outcomes of alternative policy proposals. We merely suggest a computationally more efficient way to obtain good policy proposals and to deduce the likely consequences of the proposed policies as indicated in step 11 above. Furthermore, by subjecting the econometric models to more serious scrutiny through the optimal control solution in step 8 and to continuous re-examination in step 12, it is hoped that the quality of econometric models will be improved in the process.

NOTES

1. I would like to thank Burton G. Malkiel for very helpful comments and the National Science Foundation for financial support.
2. The determination of which variables are the policy variables subject to the control of government authorities is often a difficult problem in practice. We will not discuss this issue because any policy analysis using an econometric model has already faced this issue and the purpose of our paper is to introduce stochastic control techniques to implement such policy analyses.
3. In using a planning horizon as long as 20 quarters, we are not assuming that the econometric model will be very accurate in making projections that far ahead, but we have to anticipate and incorporate the delayed effects of current policy in order to avoid recommending policies which will yield desirable results in the near future but undesirable consequences later on.

REFERENCES

G. C. Chow, *Analysis and Control of Dynamic Economic Systems* (John Wiley and Sons, 1975).

G. C. Chow, 'The Control of Nonlinear Econometric Systems with Unknown Parameters', *Econometrica*, vol. 44, no. 4 (July 1976) pp. 685–95 (a).

G. C. Chow, 'An Approach to the Feedback Control of Nonlinear Econometric Systems', *Annals of Economic and Social Measurement*, vol. 5, no. 3 (summer 1976) pp. 297–309 (b).

G. C. Chow, 'The Usefulness of Imperfect Models for the Formulation of Stabilization Policies', Econometric Research Program Research Memorandum No. 199 (Princeton University, May 1976) (c).

G. C. Chow and Sharon B. Megdal, 'An Econometric Definition of the Inflation-Unemployment Tradeoff', Econometric Research Program Research Memorandum No. 203 (Princeton University, Sep. 1976).

3 Recent Developments in Policy Optimisation for an Open Monetary Economy

John F. Helliwell

In this paper I shall outline some of the unresolved issues in the application of optimisation techniques in the choice of monetary and fiscal policies for an open economy. Many of these issues are independent of the methods used to choose policies, but most have implications for the use of optimal control techniques. Nevertheless, there is a sense in which the great efficiency of policy selection by optimisation techniques forces many of the issues into the open. Why is this? It is because any powerful optimisation technique pushes hard at all the constraints of the system, and tests all conceivable combinations of instruments. If there is a falsely loose constraint in the system, then the optimal controls will take full advantage of it. If one group of decision-makers in the model is falsely depicted as having static expectations, then the optimally-chosen policies will unmercifully exploit their gullibility. Pindyck (1973, pp. 151–3) has emphasised that this exploitive power of optimisation techniques makes them very useful in exposing '. . . characteristics of the model that would otherwise remain submerged'. Until this process is over, and there is no clear way of telling when it is over, the optimal control techniques must be regarded primarily as part of the model-building process rather than as an established procedure for policy selection.

My objective is to spell out some areas where straightforward application of optimisation techniques using established econometric models may give rise to implausible results. This sort of analysis may help to forestall problems by permitting models to be designed to avoid them, or it may merely suggest a more careful and cautious use of available models and policies.

In all of what follows, I take for granted that computing techniques are, or soon will be, available to permit the minimisation of almost any form of loss function over any arbitrary horizon subject to the constraints of any of the available range of stochastic, nonlinear, dynamic econometric models, and a set of prescribed terminal conditions.[1] I shall concentrate on the problems of policy formation in national economies that are open to substantial international flows of goods, services, capital, and population. In section I, I shall present some fairly simple propositions, and related empirical evidence, about the choice of monetary and fiscal policies under fixed exchange rate systems. The discussion in section II, relating to flexible exchange rate systems, emphasises the important role of expectations, which is the subject of section III. In section IV, I use the OPEC oil price increase as an example to illustrate some of the problems in policy selection for an open economy. Finally, in section V I list a number of model improvements that would increase the chances for successful application of econometric models in the choice of macroeconomic policies.

I POLICY CHOICES UNDER FIXED EXCHANGE RATES

Analysis of the fixed exchange rate case is important in its own right, and because it applies to some extent in all exchange systems where variations in the exchange rate are limited by official undertakings or policy preferences. Of the many propositions receiving recent attention, I offer the following selection:

1. If capital is internationally mobile, domestic credit expansion leads to offsetting capital flows that imply very large losses in foreign exchange reserves to achieve a given increase in the domestic money supply. Empirical evidence on the size of these offsets has been obtained from reduced-form estimation (e.g. Kouri and Porter, 1974) and from fully specified structural macroeconometric models (Helliwell and Lester, 1976). In the Canadian case, capital flows in the current quarter would apparently offset more than 90 per cent of any domestic credit expansion under rigidly fixed exchange rates, although the percentage drops noticeably under any system with some flexibility, even within the limited support points. The result follows for any economy in which the interest elasticity of international capital flows is high relative to the interest elasticity of the demand for money.

2. If capital is not internationally mobile, then domestic expansion may lead to increases in domestic absorption and even output, but the

induced balance of trade deficit leads to exchange reserve losses that eventually cumulate to offset fully the initial increase in domestic credit. This process, often called the 'Hume mechanism' by analogy to David Hume's analysis of the workings of the gold standard, is perhaps the central element in what has been called, by Harry Johnson (1973) and others, 'the monetary approach to the balance of payments'.

3. Neither of the above propositions vitiates the choice of monetary and fiscal policies to achieve internal and external balance as suggested by Meade (1950), Mundell (1968, pp. 152–76, 233–9, 250–71) and others. The existence of highly mobile capital narrowly constrains the movements of the domestic interest rate, but also means that foreign capital is easily available to permit, at least on a short-term basis, external balance to be consistent with sizeable trade surpluses or deficits. The operation of the Hume mechanism can be viewed as an automatic alternative to the discretionary adjustment of domestic credit to provide external balance. If the right value is chosen for domestic credit in the first place, then external balance is obtained without the Hume mechanism being called into play. If some other value is chosen, or if domestic monetary expansion is determined exogenously, say by some fixed rule, then the Hume mechanism operates (in the absence of sterilisation) to adjust the money supply, via reserve changes, to the level appropriate to external balance. The fact that the Hume mechanism works only through the cumulation of balance of payments deficits or surpluses may make it dynamically inferior to a more forward-looking monetary policy that adjusts the money supply more quickly to the level that is consistent with external balance. The choice between 'active' and 'passive' adjustment of the money supply presumably ought to depend on the nature of the disequilibrating shocks, as well as on the degree of assurance placed on the structure of the model used for policy choices.

4. If international portfolios are not growing, then the use of tight monetary and easy fiscal policies to provide internal and external balance will not be sustainable for long if it requires substantial net capital inflows on a continuing basis. For example, in the absence of portfolio growth, the inflow of capital in response to a given increase in domestic interest rates will dry up when portfolios are aligned to their newly-desired proportions. Thus continual capital inflow, unless supported by international portfolio growth, would require continual increases in interest rates. Even if there is a flow of new foreign savings to be tapped on a continuing basis, the continual increases in foreign interest payments would require successively increasing inflows of capital to attain foreign balance, as emphasised by Tsiang (1975), among others.

Thus the reliance on net capital flows to achieve internal employment and output objectives that imply a trade deficit must essentially be on a short-run basis. This suggests that proper accounting for debt service costs, and capital account equations that distinguish portfolio realignment from the distribution of new savings, are essential features of any model used for policy selection. In addition, the policy selection horizon should be long enough, or the terminal conditions strong enough, to account for or avoid the excess of domestic output over national income required to service foreign debt.

5. Increases in government expenditure may increase absorption in the short run, but in the longer run the trade balance requirement forces the eventual net effects to be zero in the simplest theoretical models with exogenous exports and with imports as a function of total absorption (e.g. Kenen, 1976, p. 50). In extremely open economies, where all output is tradeable at the world price, increases in government purchases simply absorb output that would otherwise be exported, and thus have no effect on income (McKinnon, 1975). The potential for fiscal policy to affect income even in the short run thus requires some concentration on non-traded goods, or else on expenditure that increases the scale of profitable output in the tradeable goods sector. In the longer term, it is the increase in supply potential, plus any reduction in the 'normal' import content of expenditure, that might permit government expenditure increases to lead to a sustained increase in national income and employment.

6. Prices of traded goods are set in world markets. Under fixed exchange rates, domestic inflation can exceed world inflation only to the extent that productivity increases faster in the traded goods sector than in the non-traded goods sector. This is the essence of the 'Nordic' or 'Scandinavian' approach to inflation in open economies.[2] If there are cost increases or efficiency decreases in the traded goods sector, then one does not expect, in this model, to find output priced at other than world levels. What happens, instead, is that resources gradually flow out of the traded goods sector into the non-traded goods sector where returns are greater. These shifts of resources between sectors represent one of the longer-run difficulties with fiscal policies designed, as illustrated under proposition 5, to concentrate increased government expenditure on non-traded goods so as to increase income without damaging the trade balance.

II POLICIES UNDER FLEXIBLE EXCHANGE RATES

Quite naturally, there has been a vast resurgence of academic interest in the determination of flexible exchange rates. Given this flurry of activity and the unintegrated state of the literature, the choice of a representative set of propositions is more difficult than for fixed exchange rates. I shall start without expectations and capital flows, and then add them.

1. If there are no capital flows, and the exchange rate always moves so as to achieve current account balance, then the properties of the simple open economy are very like those of a closed economy. There is no effect from foreign price shocks, as the exchange rate moves to make them invisible to the domestic economy. Terms-of-trade changes, and shifts in foreign demand at unchanged prices may have real effects on the domestic economy. Peter Kenen [1976, p. 47] argues that an increase in the demand for exports will have no effect, as the exchange rate moves to ensure that exports equal imports. However, even in the Keynesian-type model he employs there should be terms-of-trade effects that would show up if the proper distinctions were made between the price of absorption (which would be pushed down by the exchange-rate effects of the export boom) and the price of output (that would be unaffected or pushed upward). I shall return to this issue later, when I use the OPEC oil price increases as an example shock. Real effects of export shocks disappear in models that distinguish tradeable and non-tradeable goods and assume a perfectly elastic world demand for the former.

2. If capital is internationally mobile, but moves only in response to interest rate differentials (uncovered in the forward exchange market), then there are, at least in the short term, more differences between closed economies and open economies with flexible exchange rates. If capital flows so readily that the domestic interest rate is effectively pegged to the world rate, then an increase in bond-financed government expenditure can have no effect on income and employment (if terms of trade effects are ignored) because the extra bonds will all be sold to foreigners and matched by an equivalent inflow of goods and services (Mundell 1968, p. 260, Kenen 1976, p. 52). By contrast, monetary expansion (with fixed prices) is more effective than in a closed economy, because any capital outflows induced by incipient interest rate decreases will push up the price of foreign exchange and increase net exports (Mundell 1968, p. 253). By the same token, expansionary domestic monetary policy decreases income and employment abroad, because the domestic trade surplus has a counterpart trade deficit for other countries (Mundell 1968, p. 269). The foregoing propositions about the operation of the

economy under flexible exchange rates relate to theoretical models in which expectations do not appear, and which emphasise the role of the exchange rate in equating current trade and capital flows. Dornbusch (1976, p. 259) has called this a 'tradeable funds' approach to the determination of exchange rates, and notes that it ignores the important role of the exchange rate as a relative asset price. Most empirical and recent theoretical research has focused more attention on the asset-pricing role of the exchange rate.

3. Empirical research on the Canadian experience with flexible exchange rates in the 1950s and since 1970 has illustrated that fiscal expansion does have larger short-term domestic effects on real output under fixed exchange rates (with sterilisation) than under flexible exchange rates, and that domestic monetary expansion has larger real income and price effects under flexible than under fixed exchange rates.[3] This qualitative pattern does support the main elements of the two theoretical propositions reported above. However, the differences between the fixed and flexible systems are not as great as the propositions would suggest. This is because the empirical models almost all permit expectations about the future asset value of the Canadian dollar to play an important role in determining the current price of foreign exchange. If exchange rate expectations are stabilising, as the evidence suggests they were for the Canadian dollar during the 1950s and the early 1970s, then the exchange rate does not move far enough, in the short term, to make the sum of the current balance and interest-induced capital flows equal to zero. Thus expectation-induced capital movements (usually supplemented, in the Canadian case, by investment-induced capital movements) serve to release the empirical results from the more rigid of the theoretical propositions based solely on the role of the exchange rate as a device to balance current flow demands for trade and interest-induced capital flows. Not only are the effects of fiscal policy non-zero, but the proposition about money supply increases having expansionary effects at home and depressive effects abroad is rejected in the case of US monetary expansion, which apparently causes expansion in Canada as well as in the United States (Helliwell, 1974, pp. 261–2), a result that follows quite easily (but not of necessity) if exchange rate expectations are relatively sluggish in response to shocks.

4. Expectations about future exchange value, which play an obviously important role in the empirical results I have quoted, have also been at the centre of the emerging (or re-emerging, as Jacob Frenkel (1976) argues) 'asset approach' to exchange rate determination. The authors and assumptions are closely related to those of the monetary approach to

the balance of payments under fixed exchange rates. Pentti Kouri (1976, p. 281) notes that the new emphasis on the asset stock equilibrium approach, as opposed to a flow approach, to exchange rate determination reflects an empirical presumption that transaction costs are low enough to permit instantaneous portfolio adjustment. Most of the recent papers adopting an asset view of exchange rate determination accept or assume that transactions costs are low, that expectations are single-valued and 'rational', in the sense that they are formed as though the model used for analysis were an unbiased and certain predictor of the future. Some of the models (e.g. Dornbusch, 1976; Kouri, 1976) use adaptive expectations as an alternative to rational expectations, but usually to show that the use of adaptive expectations will in general ignore some of the available information and hence leave unexploited potential for arbitrage profits. In the presumed absence of substantial transactions costs, it is thus 'irrational' not to employ 'rational' expectations. Since the rationality of 'rational' expectations is conditional upon the truth of the model used for analysis, it might be preferable to follow Alan Walters' (1971) lead and call them 'consistent' expectations, to emphasise the consistency between the expectations process and the model being used for analysis.

The shorter the investment horizon, and the greater the breadth of the forward market, the more acceptable and verifiable are the assumptions about low transactions costs and instantaneous portfolio adjustment. Thus several recent papers make use of the covered interest parity relationship, whereby asset mobility ensures that the domestic interest rate minus the foreign interest rate equals the forward premium on foreign exchange. Within this framework, perfect capital mobility links domestic and foreign interest rates via the forward exchange premium or discount. If there is a monetary disturbance, say an increase in the money supply, and *if* the forward exchange rate is not affected by the change, then the spot price of foreign exchange will rise by more than the proportionate increase in the money supply (if the interest elasticity of the demand for money is less than the proportionate rate of interest; Dornbusch, 1976, pp. 261–2). If the price of forward exchange does rise, whether adapting to the current spot price or to reflect the longer-term price effects of the monetary expansion, then the spot price must move even further. Under these assumptions, with immediate adjustment of asset prices but not of domestic prices and trade flows, the effect of monetary expansion is to lead to a more than proportionate increase in the spot price of foreign exchange, an increase that will be even larger if the forward price also moves.

The foregoing example is but one of several available, most of which emphasise, by analogy with other asset markets, that any changes in market information can lead to sharp changes in the valuation of a currency without there being any changes in recorded trade and capital flows. This may seem to be in contrast to the empirical evidence reported previously, but it is all of a piece. The Canadian evidence suggests that current trade and capital flows are 'noisy' measures of the extent to which the current exchange rate deviates from its equilibrium value. Thus excess flow demands for foreign exchange lead only slowly to changes in the expected asset value of the Canadian dollar. However, and this is usually where the estimated equations erupt in a flurry of dummy variables, any time there is a rapid and widely-shared change of opinion about the value of a currency, then changes in valuation (or in reserves, to the extent that the exchange rate has official support) can take place very quickly. The slow response to the accumulation of filtered information, and the fast response to widely-shared changes in expectations are both equally at home within an asset approach to exchange rate determination. The difficult thing, and this is where the recent theoretical papers have had little to contribute, is to model the processes by which information is accumulated, and to have a theory that explains why some events and not others become triggers for sharp changes in expectations. Mussa (1976, p. 246) argues, using the presumption of rational expectations, that market participants will make the most efficient use of available information, and will change their methods if underlying circumstances, including the basic characteristics of monetary and fiscal policy, should change. The link between policies and expectations is an important one which I shall pursue later; but how are we to tell when the underlying circumstances are changing? Mussa notes that the Canadian and US dollars have been within 10 per cent of one another for 100 years, thus justifying the regressive expectations that characterise the estimated equations. If this range was breached, he argues, then the expectations processes would change. I agree, but how are we to explain the rapid drop in the value of the Canadian dollar, in terms of US dollars, from 1.03 at the end of October 1976 to .96 at the end of November, back up to 1.00 at the end of the year and back down to about .98 at the end of January 1977? Monthly foreign exchange rate equations will not be able to explain these massive changes adequately. If official purchases and sales of foreign exchange are included as a separate variable (as in the current RDX2 equation), the coefficient will suggest that the large exchange rate changes were due to perverse expectational response to official support, because the official reserves dropped by $700 million in

November and rose by the same amount during December. The official purchases and sales were no doubt intended to dampen the speed of the exchange rate change, and may or may not have done so; but we shall still need something to explain the whipsaw in expectations. The initial plunge is said to have been fuelled by the Parti Québecois victory in Quebec and the more rapid rise of labour costs in Canada than in the United States, and to have been triggered by a New York stockbroker's prediction of an 85¢ Canadian dollar. If so, then what is the source of the almost equally rapid resurgence? The suspicion must remain that even the great breadth of the foreign exchange market does not protect it from speculative surges and overshooting not founded on rational expectations about the longer-term value of the currency. These overshoots that are based solely on changes in expectations are quite different from, and more difficult to explain, than the short-run overshooting of the exchange rate that occurs in several recent theoretical papers (e.g. Dornbusch, 1976; Kouri, 1976) when the economy is digesting a monetary expansion while maintaining covered interest parity with the rest of the world.

What should be taken away from this discursive review of the asset approach to exchange rate determination, and from recent Canadian, UK, and other recent experience of exchange rate volatility? The asset approach emphasises the role of asset valuation, and raises the possibility that exchange rates can move rapidly. Experience with flexible exchange rates during the 1970s confirms the possibility of rapid expectation-induced exchange rate changes, but raises some questions about how to explain them.

III CONSISTENT EXPECTATIONS AND POLICY OPTIMISATION

There are several issues here.

1. Under certain classical assumptions, including 'rational' or 'consistent' expectations, there is no room for any effective countercyclical policy. Thomas Sargent (1976) has estimated a closed-economy model of the United States embodying these assumptions, and argues that it would not be appropriate to apply policy optimisation rules unless his classical model could be empirically rejected in favour of models that do provide scope for countercyclical policies. It is important to note, as Shiller (1975, p. 7) has emphasised, that 'the characteristic assumption made in many rational expectations macroeconomic models that variations in aggregate economic activity are due primarily to errors

made by economic agents in forecasting prices has no necessary connection to the methodology implied by the concept of rational expectations.' The connection exists because rational expectations and a classical economy are both necessary if there is to be no scope for active policies. If expectations are not consistent (and costlessly formed) then there could be room left for policy optimisation. For example, Franklin Shupp (1976) has recently illustrated how optimal control procedures could be used to define a set of wage and price controls intended to minimise the costs of moving from a current position of high expected inflation to the equilibrium rate consistent with the natural rate of unemployment. However, if private agents were as able as governments to deduce the equilibrium, and to recognise the inconsistent nature of their currently expected high rates of inflation, then there would be no scope for governments to smooth the approach to equilibrium.

2. Whatever the nature of the economy, it is apparent that private sector behaviour depends on expectations about, among other things, the policies to be adopted by governments. If policy choices come to be based on the optimisation of some specified preference function subject to an econometric model of private sector behaviour, then this new procedure will change private sector expectations and hence invalidate the econometric model. As Westcott *et al.* (1976), Bray (1975) and others have noted, a closed-loop control system with continual re-estimation of behavioural parameters might help to stop progressive decay in the quality of the policies, but there is no guarantee that it should do so. Data sufficient to estimate the current expectations processes accumulate only slowly and even then may provide only scant guide to what the expectations processes may be when current and future policies are having their effects.

3. Given the obvious importance of exchange rate expectations in open economies, governments have often attempted to improve economic performance by influencing expectations directly. What is the scope for such action? If a fixed exchange rate comes under attack, the government must seem ready to defend it until the decision to change the rate is made and implemented. This has probably come to mean that official statements about exchange rate policy intentions are ignored or even treated as intentionally misleading. Under flexible exchange rates the situation is less clear. When the Canadian government expressed a clear intention in June 1961 to use its policies to achieve a lower value for the Canadian dollar, the exchange rate did move down fairly rapidly from 1.01 ($US/$Can.) to about .97. However, when the government started to use its foreign exchange reserves in early 1962 to stop the rate

going below .95, a large speculative outflow ensued. The 'flexible' exchange rate was abandoned in May 1971 in favour of a rate pegged at .925. Even that rate was subsequently 'attacked' and successfully 'defended' by domestic 'austerity' and international credit until 'confidence' was re-established.

It is notoriously difficult to disentangle the expectations effects of pronouncements and performance. Within any framework of consistent expectations, one ought to expect that the credibility of official pronouncements will depend on the past record of reliability. However, governments speak with so many different voices, their participants and policies change so quickly, and they take so little care to distinguish policy intentions, conditional forecasts, and unconditional forecasts, that the hypothesis would be extremely difficult to test. Why does the issue matter? It matters because sometimes the success of a policy depends on the ability to establish expectations that are consistent with the policy. For example, a policy of using the exchange rate to insulate the domestic economy from foreign inflation can only work smoothly if private expectations are consistent with the policy. The Economic Council of Canada has recommended, in its last three Annual Reviews, that Canada adopt a domestic inflation target equal to $\pm\frac{1}{2}$ per cent of a weighted average of the consumer price indices of Canada's main trading partners. The only apparent rationale for such a policy is that private expectations would not let the exchange rate move enough to cushion Canada from foreign inflation. As a consequence, world inflation unaccompanied by matching Canadian inflation would lead either to unemployment in the non-traded goods sector (as domestic demand switched to traded goods) or, in a fully employed economy, to potentially costly adjustments as resources move from the non-traded to the traded goods industries in response to the higher profits in the latter (Dornbusch, 1976, pp. 265–7). These problems could be reduced or eliminated only if the exchange rate were expected to move enough to absorb the difference between Canadian target price levels and foreign price levels. This would involve a substantial change from the existing policy framework, wherein domestic inflation has been attributed to foreign inflation, and the Economic Council has been advocating that this link should be officially strengthened. If such a change were contemplated, it could only overcome the effects of inelastic private sector expectations if it were also possible for the government's intention to pursue an independent inflation target to be announced and believed.

The three issues raised in this section can be summarised. First, if expectations are rational, then the scope for helpful policies is reduced, especially if the economy tends to full employment in the absence of disturbances. Second, whatever the nature of the expectations processes, subjecting the economy to optimal control will generally alter the structure of expectations and hence will bias the control process. Third, some policy changes require changes in expectations processes if they are to be effective. Expectations that are consistent with the proposed policy may not be consistent with past experience. In the short term, governments have an incentive to influence private expectations directly, even at the expense of those who follow the advice. In the longer term, this entails a loss of credibility and makes it more difficult for policy-makers to alter private expectations even when the alteration is in the interest of all parties.

IV THE MACROECONOMICS OF OIL PRICE INCREASES

The 1972–7 experience of combined inflation, lost output, and unemployment has been clearly worse than any period in the preceding twenty-five years. If a strong case is to be made that policy optimisation is possible through the use of optimal control in conjunction with econometric models, then one ought to be able to show that better policy responses were possible, and that existing methods and models would have helped to find them. I shall consider briefly how alternative macroeconomic models would have depicted the effects of a large oil price increase. I do not pretend that the oil price increase was an isolated exogenous event, or that there would not have been repercussions from the 1971–3 boom even if OPEC had not acted so vigorously. I use the oil price increase as an example shock in part because it is difficult for models to handle, and in part because any model that could not deal with the effects of a terms-of-trade change could not be an adequate guide to policy-making for an open economy.

I shall deal only with flexible exchange rates, starting first without capital flows and then adding them. The focus of analysis will be a country that is a major importer of OPEC oil, the price of which has been substantially increased. Immediately the economy suffers an inflationary shock and an incipient trade deficit. Given the higher prices and incipient deficit, a simple-minded policy response might be to tighten monetary and fiscal policies so as to reduce absorption and prices. When added to the depressive real balance and wealth effects of the oil-price increase, this policy would be risky and possibly counterproductive for a single country, and collectively doomed to failure if applied by all the non-

OPEC countries. The goal of such a policy would be to reduce non-OPEC income and prices far enough that OPEC would not have a current-account surplus. Even for individual countries, and certainly for groups of countries, facing a terms-of-trade shift where short-term price elasticities are small, the OPEC example suggests that the exchange rate cannot by itself provide appropriate insulation by balancing the trade account.

If there are interest-induced capital flows, then there is a broader range of policy options for dealing with terms-of-trade changes. Returning to the OPEC example, the inevitability of OPEC current-account surpluses implies new additions to the portfolio of investible funds. From the point of view of policy optimisation, the problem for each country was and is to choose the appropriate current account deficit and capital account surplus to best cushion domestic prices and output from the OPEC shock. Westcott *et al.* (1976, p. 23) note that the sheer size of the OPEC price shock would have made suspect the use of an optimal control rule based on experience over the preceding fifteen years. I think a stronger point can be made: that any feedback rule determining policies in the light only of internal (price, output, and employment) and external balance (however measured) targets could not adequately respond to a terms-of-trade shock. Any appropriate response must pay due heed to the source of the shock so as to be able to choose a preferred path of adjustment. This leaves open the obvious problem of identifying systematic shocks and isolating them from the random noise that is regularly embodied in macroeconomic applications of stochastic optimal control (e. g. Bray, 1975).

I have suggested that the standard rules for policy choice in national economies would not have helped much in digesting the oil price increases. Would the classical monetarist prescriptions have fared much better? Marcus Miller (1976, pp. 507–10) has argued that the monetarist policy would have been more deflationary than any of the alternatives. On the classical assumptions of full employment and costless adjustment, coupled with rational expectations, an unchanged world money supply would have caused other prices to drop, and permitted real incomes and wealth to be redistributed without loss in output. On this rather extreme view, any drop in real output that did take place must be attributed either to mistaken policies or to a temporary failure of rational expectations. In this vein David Laidler (1976) lays the main blame, in the case of the United Kingdom, on excessive public borrowing, failure to recognise an increase in the natural unemployment rate, and excessive monetary ease in 1972–3 followed by monetary tightness in 1973–74. Rather surprisingly, his explanation of UK inflation and unemployment since 1974 attaches no importance to external factors, perhaps because the flexible exchange rate is supposed to provide adequate insulation, or because Laidler was objecting to other views that attempt to place the blame entirely on external factors. If one accepts modifications to the classical assumptions sufficient to allow domestic monetary policy to have effects on domestic real output, then terms of trade changes ought likewise to have real effects.

To the best of my knowledge, none of the existing national econometric models, whatever their mix of Keynesian and classical assumptions, have structures that depict adequately the macroeconomic effects of a terms-of-trade change. Even if there had been more appropriate national models on hand, attempts to use such models as constraints in the choice of policies in response to the oil-price increase (even on the false assumption that the oil price increase was entirely exogenous) would almost surely have led to the globally self-defeating results that I outlined earlier. Only by using a fairly long policy horizon and by international coordination could national models have produced helpful and consistent policy suggestions.

V FUTURE REQUIREMENTS FOR OPEN ECONOMY MODELS

To close my discursive and somewhat sceptical survey on a positive note, I shall briefly list some of the features that national models are likely to need if they are to be of much use in improving policy choices for open economies.

1. The productive potential of the economy needs much more secure modelling, taking due account of the productivity of private and public investment.

2. The microeconomics of the government sector need to be more closely specified, to give a better idea of which expenditures and taxes, if any, can be best used for countercyclical purposes.

3. What can be learned about the pattern of the so-called 'random disturbances' that are supposed to drive most stochastic models? Which ones can and ought to be countered as they hit the economy, and which ones are the harbingers of a new order to which the economy must adjust?

4. More generally, are there ways of assessing the links between efficiency and stability in the macroeconomy? If these links exist, then they should affect the policy optimisation procedure.

5. More effort needs to be devoted to establishing the costs and lags involved in changing values of policy variables.

6. Much better modelling is required to measure the degree of openness to international movements of population, labour force, services, and capital. Research on the Canadian case[4] suggests that international transmission and domestic responses to policy changes and international disturbances are materially different if migration is taken into account. Similar linkages within Europe are likely to be even stronger. Special care should be attached to estimating trends in the degree of openness, to guard against the use of out-of-date equations to evaluate forward-looking policies.

7. More attention to the open-economy aspects of national models suggests a parallel increase in policy studies in an international context. Project LINK provides a valuable forum for such studies, as do the linked models of European countries, but additional efforts should be encouraged.

8. Model diversity should in general be encouraged, so as to permit better understanding and testing of alternative structures. Existing methods for optimal policy choice within a stochastic environment can fairly easily accommodate stochastic error terms, stochastic exogenous variables, and even uncertainty about coefficient estimates, but the model structure itself is always taken as given. This has the effect of heavily understating the uncertainty of policy assessments, as differences between models, and even between different versions of the same model, can give vast differences in model results. One disturbing aspect of the disparity of properties of even quite similar model structures is that the alternative models are often very similar in their goodness of fit to historical data. By the time sufficient out-of-sample data have accumulated to permit more clear-cut preferences among alternative models, all of the structures have become somewhat dated.

NOTES

1. I am not suggesting that the optimisation processes are or will become trivially cheap, despite continuing decreases in the real cost of computing. Optimisation using large non-linear models will always require a lot of computing. For example, Robert Holbrook (1973, p. 70) notes that 1440 solutions of RDX2 (each of which may require from 10 to 30 iterations to solve the 260-odd equations of the model) are need to find the optimal values for four instruments over an eight-quarter horizon.
2. A primary source is Edgren, Faxen, and Odhner (1969). Earlier work emphasising the split between traded and non-traded goods includes Salter

(1959). Recent applications in a monetary context include Paunio and Halttunen (1976), Jonson and Kierzkowski (1975) and Parkin (1977).
3. The main RDX2 results for monetary policies are reported in Helliwell and Maxwell (1974), and for fiscal policies in Helliwell (1974). The corresponding results from other models of the Canadian economy are surveyed in Helliwell (1975).
4. The evidence is reported in Helliwell (1974).

REFERENCES

Bray, Jeremy (1975), 'Optimal Control of a Noisy Economy with the U.K. as an Example', *J. R. Statist. Soc.*, A, part 3, *138*, pp. 339–66.

Dornbusch, Rudiger (1976), 'The Theory of Flexible Exchange Rate Regimes and Macroeconomic Policy', *Scandin. J. Econ.*, no. 2, *78*, pp. 255–75.

Edgren, Gösta, K. O. Faxen and C. E. Odhner (1969), 'Wages, Growth, and the Distribution of Income', *Swed. J. Econ.*, no. 3, *71*, pp. 133–60.

Frenkel, Jacob A. (1976), 'A Monetary Approach to the Exchange Rate: Doctrinal Aspects and Empirical Evidence', *Scandin. J. Econ.*, no. 2, *78*, pp. 200–24.

Helliwell, John F. (1974), 'Trade, Capital Flows, and Migration as Channels for International Transmission of Stabilization Policies', in A. Ando, R. Herring, and R. Marston, eds, *International Aspects of Stabilization Policies* (Federal Reserve Bank of Boston, Boston, 1974) pp. 241 78.

Helliwell, John F. (1975), 'Adjustment Under Fixed and Flexible Exchange Rates', in P. B. Kenen, ed. *International Trade and Finance: Frontiers for Research* (Cambridge U.P., London) pp. 379–410.

Helliwell, John, and Tom Maxwell (1974), 'Monetary Interdependence of Canada and the United States under Alternative Exchange Rate Systems', in R. Z. Aliber, ed. *National Monetary Policies and the International Financial System* (University of Chicago Press) pp. 82–108.

Helliwell, John F., and John M. Lester (1976), 'External Linkages of the Canadian Monetary System', *Can. J. Econ.*, Nov 1976, *9*, pp. 648–69.

Holbrook, Robert S. (1973), *An Approach to the Choice of Optimal Policy using Large Econometric Models* (Bank of Canada, Ottawa: Staff Research Studies, no. 8).

Johnson, Harry G. (1973), 'The Monetary Approach to Balance-of-Payments Theory', in M. B. Connolly and A. K. Swoboda, eds, *International Trade and Money: The Geneva Essays* (Allen & Unwin, London) pp. 206–24.

Jonson, P. D., and H. Kierzkowski (1975), 'The Balance of Payments: An Analytic Exercise', *Manch. Sch.*, June 1975, pp. 105–34.

Kenen, Peter B. (1976), *Capital Mobility and Financial Integration: A Survey* (International Finance Section, Princeton Univ.: Princeton Studies in International Finance no. 39).

Kouri, Pentti J. K. (1976), 'The Exchange Rate in the Short Run and in the Long Run: A Monetary Approach', *Scandin. J. Econ.*, no. 2, 78, pp. 280–304.

Kouri, P. J. K., and M. G. Porter (1974), 'International Capital Flows and Portfolio Equilibrium', *J. Polit. Econ.*, May/June, *82*, pp. 443–68.

Laidler, David (1976), 'Inflation in Britain: A Monetarist Perspective', *Amer. Econ. Rev.*, Sep, *66*, pp. 485–500.

McKinnon, Ronald I. (1975), 'The Limited Role of Fiscal Policy in an Open Economy', paper prepared for a conference on the Monetary Mechanism in Open Economies sponsored by the Bank of Finland, Helsinki, Aug 1975.

Meade, James E. (1951), *Balance of Payments* (Oxford University Press, London).

Miller, Marcus H. (1976), 'Can a Rise in Import Prices be Inflationary and Deflationary? Economists and the U.K. Inflation, 1973–74', *Amer. Econ. Rev.*, Sep, 66, pp. 501–19.

Mundell, Robert A. (1968), *International Economics* (Macmillan and Collier-Macmillan, New York and London).

Mussa, Michael (1976), 'The Exchange Rate, the Balance of Payments and Monetary and Fiscal Policy Under a Regime of Controlled Floating', *Scandin. J. Econ.*, no. 2, 78, pp. 229–48.

Parkin, J. Michael (1977), 'World Inflation, International Relative Prices, and Monetary Equilibrium under Fixed Exchange Rates', in R. Z. Aliber, ed, *The Political Economy of Monetary Reform* (Macmillan, London).

Paunio, J. J., and H. Halttunen (1976), 'The "Nordic" Approach to Inflation: Interpretation and Comments', in J. M. Parkin and M. Zis, eds, *Inflation in the World Economy* (Manchester Univ. Press) pp. 259–70.

Pindyck, Robert S. (1973), *Optimal Planning for Economic Stabilization* (North-Holland, Amsterdam).

Salter, W. E. G. (1959), 'Internal and External Balance: The Role of Price and Expenditure Effects', *Econ. Rec.*, Aug, *35*, pp. 226–38.

Sargent, Thomas J. (1976), 'A Classical Macroeconometric Model for the United States', *J. Polit. Econ.*, Apr, *84*, pp. 207–38.

Shiller, Robert J. (1975), 'Rational Expectations and the Dynamic Structure of Macroeconomic Models', paper prepared for a conference on the Monetary Mechanism in Open Economies, sponsored by the Bank of Finland, Helsinki, Aug 1975.

Shupp, Frankin R. (1976), 'Optimal Policy Rules for a Temporary Incomes Policy', *Rev. Econ. Stud.*, June, *43*, pp. 249–60.

Tsiang, S. C. (1975), 'The Dynamics of International Capital Flows and Internal and External Balance', *Quart. J. Econ.*, May, *89*, pp. 195–214.

Walters, A. A. (1971), 'Consistent Expectations, Distributed Lags and the Quantity Theory', *Econ. J.*, June, pp. 273–81.

Westcott, J. H., S. Holly, B. Rüstem, and M. B. Zarrop, 'Control Theory in the Formulation of Economic Policy' (Imperial College, London, PREM Discussion Paper no. 17: Submission to the Committee of Inquiry on Policy Optimisation).

4 The Role of Feedback in Macroeconomic Policy

D. A. Livesey

Whilst economists have been exploring the application of optimal control theory to macroeconomic policy formulation some engineers have been re-examining classical feedback theory. Recently the feedback literature has been extended to cover multivariable systems in a way which gives insight into questions of stability and control. In this paper some of the features of this feedback approach most relevant to economic applications are demonstrated using two different macroeconomic models. We begin with a version of Bill Phillips' model which he used in his famous article (Phillips (1954)), to illustrate the application of control theory to economies. Then we turn to Wassily Leontief's dynamic input-output model which is a large linear model used for planning purposes in many countries. Finally we examine how the feedback techniques could be used to analyse a nonlinear model.

I A SMALL LINEAR MODEL

Feedback theory has always been concerned with the relationship between a system's inputs and its outputs and, in particular, with the way in which the system changes when the inputs are varied in response to information available about the outputs. A system's characteristic behaviour is entirely determined by the transfer function which maps the inputs into the outputs. It was, therefore, entirely appropriate that the classical feedback literature concentrated on the analysis of transfer functions. More recent work has taken as its starting-point the familiar state-space representation of a linear model.

$$\begin{aligned} x(t+1) &= Fx(t) + Gu(t) \\ y(t) &= Cx(t) + Du(t) \end{aligned} \tag{1.1}$$

The control variables, $u(t)$, may be a function either of the states, $x(t)$, or of the outputs, $y(t)$. Suppose

$$u(t) = kKy(t), \tag{1.2}$$

where k is a scalar gain and K is the feedback matrix. Then

$$x(t+1) = [F + kG(I - kDK)^{-1}C]x(t), \tag{1.3}$$

determines the system's behaviour when there is feedback. Note that when $k = 0$, i.e. no feedback, the matrix F, the state transition matrix, completely determines the system's response. In this short paper we cannot hope to present any of the details of the recent advances in multivariable feedback theory—the reader should consult MacFarlane (1975) for this. Here we demonstrate some of the features of the Cambridge Linear Analysis and Design Program and I am grateful to Alistair MacFarlane and John Edmunds for their encouragement to explore the package's potentials. Any errors and misrepresentations are my sole responsibility.

There is nothing particularly novel about the multiplier-accelerator model used by Phillips in his (1954) paper on the stabilisation of economic policy. It now seems to be an established tradition to use it for illustrative purposes in papers on linear decision rule theory. Phillips specified a continuous-time model, whereas here we shall use a discrete-time version; and, since it comes in various forms depending upon which author one consults, we use our own notation.

$$E = C + I + D + B \tag{1.4}$$

This is an accounting identity equating aggregate demand, E, to the sum of consumption, C, investment, I, autonomous injections, D, and the balance of trade, B. Consumption is treated as being proportional to aggregate production, P;

$$C = (1 - s)P + \varepsilon_c: \tag{1.5}$$

investment is a lagged proportional response to changes in aggregate production;

$$I(t) = \kappa v[P(t) - P(t-1)] + (1 - \kappa)I(t-1) + \varepsilon_i(t): \tag{1.6}$$

and production is assumed to lag on demand.

$$P(t) = \alpha E(t) + (1 - \alpha)P(t-1) + \varepsilon_p(t). \tag{1.7}$$

We also assume that autonomous injections are made up of not only government expenditure, G, which is a policy variable, but also a disturbance term, ε_g;

$$D = G + \varepsilon_g. \tag{1.8}$$

Phillips' model can be extended to include foreign trade if we assume that exports, X, are given by

$$X = xP + \varepsilon_x, \tag{1.9}$$

and imports, M, are determined by

$$M = mP + \theta(E - P). \tag{1.10}$$

Both of these equations come from Aoki (1973). If $x = 0$ we have the more usual assumption that exports are exogenously determined by the disturbance term, ε_x. Equation (1.10) implies that imports are determined by both the level of production and excess demand. The balance of trade is

$$B = X - M, \tag{1.11}$$

exports less imports. We might have chosen to include a disturbance term in the import equation. It is obvious however that the model would only be affected by the difference between the disturbance terms in the import and export equations. Thus ε_x adequately represents the most general case.

Since equations (1.5) to (1.10) are *behavioural* equations which provide a theoretical explanation of empirically observable phenomena, it is assumed that not only the parameters, s, α, etc., can be estimated but also the error terms, ε_c, ε_i, etc. Here we have little to say about the disturbance terms and we will ignore them all except ε_c which will be useful later.

A state space representation of the model is

$$\begin{aligned} x^*(t+1) = {} & \frac{1}{\Delta}\begin{bmatrix} \Delta + \alpha\psi & \alpha(1-\kappa) \\ \alpha\kappa\nu\psi & (1-\kappa)(\Delta + \alpha\kappa\nu) \end{bmatrix} x^*(t) \\ & + \frac{\alpha}{\Delta^2}\begin{bmatrix} (1-\alpha)(1+\theta) - \alpha\kappa^2\nu \\ (1-\kappa)\kappa\nu(1-\alpha)(1+\theta) + \alpha\kappa^2\nu\psi \end{bmatrix} u(t) \\ & + \frac{\alpha}{\Delta^2}\begin{bmatrix} (1-\alpha)(1+\theta) - \alpha\kappa^2\nu \\ (1-\kappa)\kappa\nu(1-\alpha)(1+\theta) + \alpha\kappa^2\nu\psi \end{bmatrix} \varepsilon_c \end{aligned} \tag{1.12}$$

and the observation equation is

$$y(t) = \begin{bmatrix} P(t) \\ I(t) \\ E(t) \\ C(t) \\ B(t) \\ D(t) \\ X(t) \\ M(t) \end{bmatrix} = \begin{bmatrix} 1 & 0 \\ 0 & 1 \\ 1+\psi/1+\theta & 1/1+\theta \\ 1-s & 0 \\ x-m-\psi\theta/1+\theta & -\theta/1+\theta \\ 0 & 0 \\ x & 0 \\ m+\psi\theta/1+\theta & \theta/1+\theta \end{bmatrix} [x^*(t)]$$

$$+\frac{1}{\Delta}\begin{bmatrix} \alpha \\ \\ \alpha\kappa v \\ 1 \\ \alpha(1-s) \\ \pi \\ \Delta \\ \alpha x \\ \alpha x - \pi \end{bmatrix} G(t) + \frac{1}{\Delta}\begin{bmatrix} \alpha \\ \\ \alpha\kappa v \\ 1 \\ \alpha(1-s) \\ \pi \\ \Delta \\ \alpha x \\ \alpha x - \pi \end{bmatrix} \varepsilon_c \tag{1.13}$$

We have defined

$$\Delta = (1-\alpha)(1+\theta) - \alpha\psi - \alpha\kappa v, \; \psi = x - m - s,$$

and

$$\pi = (x-m)\alpha - \theta(1-\alpha).$$

For this model the final government expenditures multiplier value for production is $1/\psi$.

In order to demonstrate the elements of feedback theory, we need to use a numerical example. Setting $\alpha = 0.8$, $s = 0.3$, $v = 0.6$ and $\kappa = 0.5$, we are using Phillips' original values, in addition we assume that $x = 0.25$, $m = 0.2$, and $\theta = 0$. Ignoring the disturbance term, we get the state equation,

$$x(t+1) = \begin{bmatrix} -0.25 & 2.5 \\ -0.375 & 1.25 \end{bmatrix} x(t) + \begin{bmatrix} 2.5 \\ 0 \end{bmatrix} G(t). \tag{1.14}$$

If we assume that we only have data available for P and I then the observation equation is

$$y(t) = \begin{bmatrix} P(t) \\ I(t) \end{bmatrix} = \begin{bmatrix} 1 & 0 \\ 0 & 1 \end{bmatrix} x(t) + \begin{bmatrix} 5 \\ 1.5 \end{bmatrix} G(t). \tag{1.15}$$

Suppose that we decide to control the economy using the rule

$$G(t) = -kP(t), \tag{1.16}$$

where k is a scalar gain parameter. We would be changing government expenditure in response to changes in production in the standard Keynesian way. Intuitively, we might decide to set k equal to the reciprocal of the multiplier; i.e. in our model, $\psi = 0.25$. In Figure 4.1 we have the root loci for the Phillips model as the gain k increases from 0.001 upwards. We note at once that we are always within the unit circle even for very high values of gain. In other words with these particular parameter values the model is very stable.

One source of instability in the economy which we have not captured in the model so far are the delays in gathering data. Suppose we have to

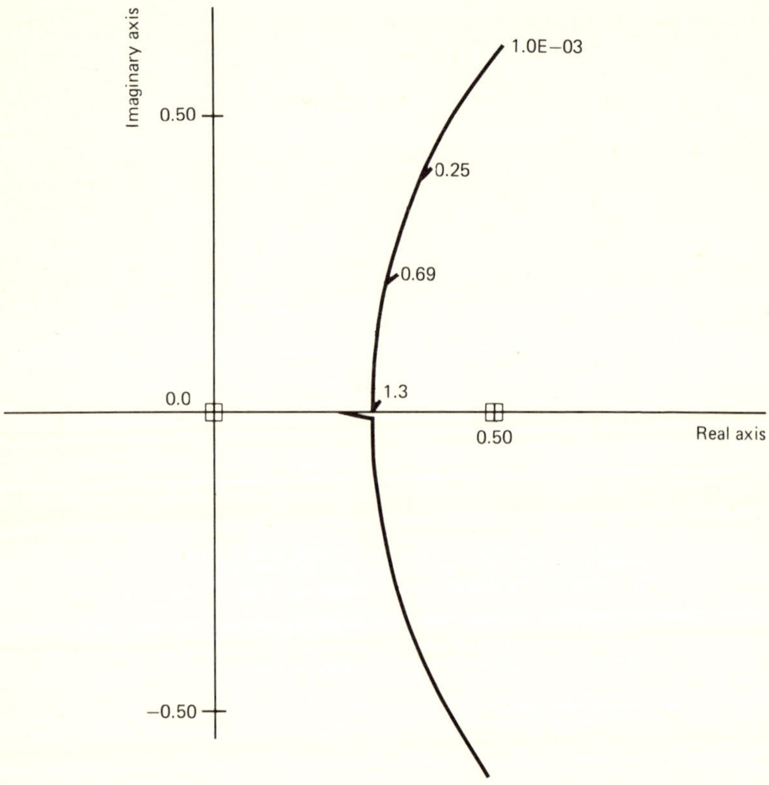

FIG. 4.1 Phillips Model with Negative Feedback

wait one period before our observations of P and I are available. The model then becomes:

$$x(t+1) = \begin{bmatrix} -0.25 & 2.5 & 0 & 0 \\ -0.375 & 1.25 & 0 & 0 \\ 1 & 0 & 0 & 0 \\ 0 & 1 & 0 & 0 \end{bmatrix} x(t) + \begin{bmatrix} 2.5 \\ 0 \\ 5 \\ 1.5 \end{bmatrix} G(t)$$

$$y(t) = \begin{bmatrix} P(t-1) \\ I(t-1) \end{bmatrix} = \begin{bmatrix} 0 & 0 & 1 & 0 \\ 0 & 0 & 0 & 1 \end{bmatrix} x(t). \qquad (1.17)$$

If we now apply the rule

$$G(t) = -kP(t-1), \qquad (1.18)$$

we might expect the model to be more unstable. Figure 4.2 shows this to be the case when there are high values of the gain, k. Note, however, that

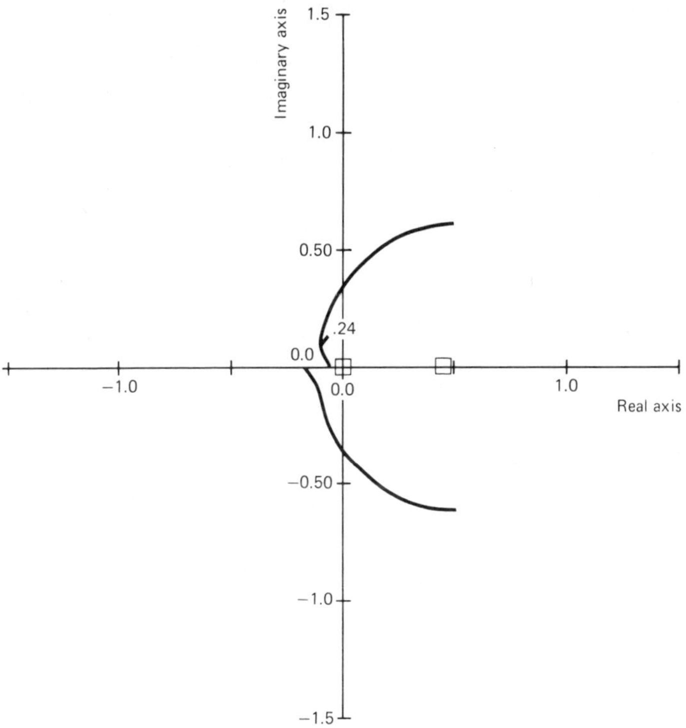

FIG. 4.2 With Time Delays

when $k = 0.25$ we have a very stable and non-oscillatory system. This is, the gain we intuitively derived using the theory of the multiplier.

If the information delays were longer than one period we would expect the root loci to be even further to the left and the system correspondingly more unstable.

We can use feedback theory to examine the sensitivity of the model to changes in its parameters. Suppose that, for the time being, we set $s = 1$, then

$$C = \varepsilon_c.$$

If we assume that government expenditure is fixed and not a control variable then the model becomes

$$x(t+1) = \frac{1}{0.9}\begin{bmatrix} -0.05 & 0.5 \\ -0.285 & 0.6 \end{bmatrix} x(t) + \frac{1}{0.9^2}\begin{bmatrix} 0.1 \\ -0.105 \end{bmatrix} \varepsilon_c(t)$$

and

$$y(t) = \begin{bmatrix} P(t) \\ I(t) \end{bmatrix} = \begin{bmatrix} 1 & 0 \\ 0 & 1 \end{bmatrix} x(t) + \frac{1}{0.9}\begin{bmatrix} 1 \\ 0.3 \end{bmatrix} \varepsilon_c(t). \tag{1.19}$$

Using the feedback rule

$$\varepsilon_c(t) = kP(t)$$

is in effect to assume that

$$C = (1-s)\,P(t), \tag{1.20}$$

where

$$s = 1 - k.$$

Hence, we are able to obtain the effects of altering the parameter s in the model whilst the others remain unchanged.

Figure 4.3 shows the root loci for the Phillips model as k is increased from 0. Notice that by the time that $k = 0.8$, i.e. $s = 0.2$, the model has become unstable since the roots are outside the unit circle.

These are but a few examples of the way in which the linear feedback literature can be used to analyse economic models.

II LEONTIEF'S DYNAMIC INPUT-OUTPUT MODEL

Leontief's dynamic input-output model is one of the best-known linear models in economics. Its importance derives from its ability to

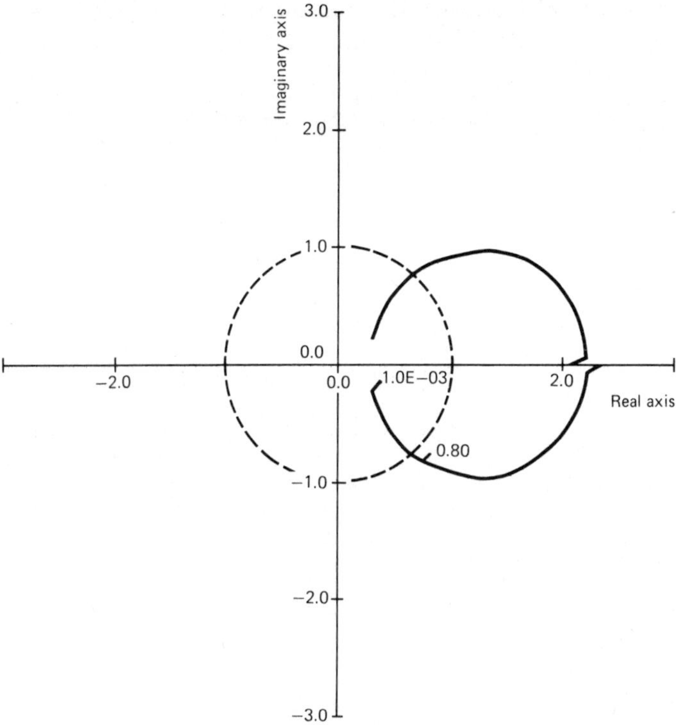

FIG. 4.3 Phillips Model with Variable Savings Propensity

disaggregate economic behaviour into a large number of industries and/or commodities. At first sight the Leontief model is almost in the state space form. There is however a major obstacle since the obvious state-space representation involves the inversion of a singular matrix. A solution to this difficulty has been proposed in Livesey (1976), which uses the concepts of realisation theory.

We are considering the multisectoral dynamical input-output model, which is normally represented by

$$\underline{q}_t = A\underline{q}_t + B(\underline{q}_{t+1} - \underline{q}_t) + f_t, \tag{2.1}$$

where

$\underline{q}_t$ = gross output levels
f_t = final demand, excluding investment
A = Leontief technical coefficient matrix
B = capital coefficient matrix.

Since the composition of final demand, excluding investment, is taken to be exogenously determined, equation (2.1) could be written, provided that the matrix B is nonsingular, as

$$\underline{q}_{t+1} = B^{-1}R\underline{q}_t - B^{-1}f_t, \tag{2.2}$$

where

$$R = I - A + B.$$

In this paper we use a four-sector model with coefficient matrices given below. These matrices are taken from Barker (1971) and we notice

A Projected intermediate inputs per unit of output, Britain 1972

	Direct inputs in 1960 £ per 1960 £ of output			
	1	*2*	*3*	*4*
Commodity	*Agriculture and food*	*Manufacturing*	*Mining, construction, utilities*	*Transport and services*
1. Agriculture and food	0.3208	0.0063	0.0199	0.0031
2. Manufacturing	0.1648	0.4602	0.2611	0.1523
3. Mining, construction and utilities	0.0361	0.0778	0.1784	0.0373
4. Transport and services	0.0886	0.0770	0.0821	0.0827

B Projected capital inputs per unit of output, Britain 1972

	Direct inputs in 1960 £ per 1960 £ of output			
	1	*2*	*3*	*4*
Commodity	*Agriculture and food*	*Manufacturing*	*Mining, construction, utilities*	*Transport and services*
1. Agriculture and food	0.0000	0.0000	0.0000	0.0000
2. Manufacturing	0.5922	0.7486	1.1393	1.3646
3. Mining, construction and utilities	0.2445	0.2456	0.9983	0.9512
4. Transport and services	0.0638	0.0766	0.1736	0.1724

immediately that the matrix B is singular. This singularity arises because no output from the agriculture and food sector is used in the production of investment goods. We can rearrange B so that the first n_1 rows are non-zero and the last $n-n_1$ rows contain all the zero rows. This allows us to partition equation (2.1) as

$$\left[\frac{\underline{q}^1}{\underline{q}^2}\right]_t = \left[\frac{A_{11} \quad A_{12}}{A_{21} \quad A_{22}}\right]\left[\frac{\underline{q}^1}{\underline{q}^2}\right]_t + \left[\frac{B_{11} \quad B_{12}}{0 \quad 0}\right]\left\{\left[\frac{\underline{q}^1}{\underline{q}^2}\right]_{t+1} - \left[\frac{\underline{q}^1}{\underline{q}^2}\right]_t\right\} + \left[\frac{\underline{f}^1}{\underline{f}^2}\right]_t. \tag{2.3}$$

The partitioning of the reordered vectors $\underline{q}_t$ and $\underline{f}_t$ reflects the fact that those commodities that do not go into capital goods appear as the elements of the vectors $\underline{q}^2$ and $\underline{f}^2$. If we define a transformation matrix

$$J_2 = \left[\frac{0 \quad 0}{0 \quad I_{n-n_1}}\right],$$

where I_k denotes the unit matrix of rank k and the partitioning corresponds to that used for the B matrix in equation (2.3), then we may write the equation (2.3) as

$$\underline{q}^*_{t+1} = E^{*-1}[I+E]\underline{q}^*_t - [E^{*-1}J_2 + I]\underline{f}_t. \tag{2.4}$$

We have defined

$$E = B(I-A)^{-1},\ B^* = B - J_2(I-A),$$

and

$$E^* = B^*(I-A)^{-1} = B(I-A)^{-1} - J_2 = E - J_2. \tag{2.5}$$

Thus

$$E = \left[\frac{E_{11}}{0}\middle|\frac{E_{12}}{0}\right] \text{ and } E^* = \left[\frac{E_{11}}{0}\middle|\frac{E_{12}}{-I}\right].$$

The observation equation corresponding to (2.4) is

$$\underline{q}_t = B^{*-1}\underline{q}^*_t - B^{*-1}J_2\underline{f}_t, \tag{2.6}$$

which defines the state variables as

$$\begin{aligned}\underline{q}_t^* &= B^*\underline{q}_t + J_2\underline{f}_t \\ &= B\underline{q}_t + J_2[\underline{f}_t - (I-A)\underline{q}_t] \\ &= B\underline{q}_t\end{aligned} \tag{2.7}$$

The final term in the penultimate line of equation (2.7) is zero. We now have an obvious definition for the state variables, since the elements of the vector $B\underline{x}_t$ are the stocks of outputs that enter the implicit production function in the dynamic input-output model.

In Livesey (1976) it is shown that (2.7) is an uncontrollable representation of the dynamic Leontief model. A valuable key is provided by equation (2.7) as to how one might best proceed. We note that writing this definition in its partitioned form,

$$\left[\begin{array}{c|c} B_{11} & B_{12} \\ \hline 0 & 0 \end{array}\right]\left[\begin{array}{c} \underline{q}_1 \\ \hline \underline{q}_2 \end{array}\right] = \underline{q}_t^* \left[\begin{array}{c} k_t \\ \hline 0 \end{array}\right], \tag{2.8}$$

defines a new vector, $\underline{k}_t$, which has n_1 elements. Each element, k_t^i, is the stock of the ith industry's output required to produce at the level of gross output, $\underline{q}_t$. That not all types of industrial output go into capital formation explains the zero rows in the B matrix and accounts for the zero elements of the vector, $\underline{q}_t^*$.

Suppose, therefore, that we define

$$\underline{k}_t = J\underline{q}_t^*, \text{where } J = [I_{n_1}|0],$$

then

$$\underline{q}_t^* = J^*\underline{k}_t, \text{ where } J^* = \left[\begin{array}{c} I_{n_1} \\ \hline 0 \end{array}\right].$$

Thus the state equation (2.4) may be transformed into

$$\begin{aligned} \underline{k}_{t+1} &= J[I + E^{*^{-1}}]\, J^*\underline{k}_t - J[E^{*^{-1}}J_2 + I]f_t \\ &= [I + E_{11}^{-1}]\underline{k}_t - [I|E_{11}^{-1}E_{12}]f_t, \end{aligned} \tag{2.9}$$

where the corresponding observation equation is

$$\underline{q}_t = B^{*^{-1}}J^*\underline{k}_t - B^{*^{-1}}J_2 f_t. \tag{2.10}$$

Equations (2.9) and (2.10) form a controllable and observable proper realisation of the Leontief dynamic input-output model.

Using Barker's estimates, given above, for the A and B matrices we obtain the state equation:

$$k_{t+1} = \begin{bmatrix} -7.32 & -10.1 & 120.69 \\ -376.94 & -301.63 & 4653.13 \\ 499.33 & 404.09 & -6175.57 \end{bmatrix} k_t$$

$$+\begin{bmatrix} -1 & 0 & 0 & -0.689 \\ 0 & -1 & 0 & 0.460 \\ 0 & 0 & -1 & -0.776 \end{bmatrix} f_t \qquad (2.11)$$

and the corresponding observation equation is:

$$\underline{q}_t = \begin{bmatrix} -94.8 & -81.8 & 1202.3 \\ -445.8 & -358.3 & 5505.5 \\ 495.3 & 400.6 & -6125.6 \\ -11.68 & -9.43 & 144.5 \end{bmatrix} k_t$$

$$+\begin{bmatrix} 0 & 0 & 0 & -0.7786 \\ 0 & 0 & 0 & 0.5177 \\ 0 & 0 & 0 & 0.7220 \\ 0 & 0 & 0 & 1.48 \end{bmatrix} f_t \qquad (2.12)$$

The coefficient matrices, particularly the state transition matrix, look fairly nasty and this is confirmed by the calculation of the open loop eigenvalues and eigenvectors. They are:

Eigenvalue	*−6489*	*1.28*	*3.99*
Corresponding eigenvector:	0.0194	0.5	0.5
	0.7401	0.2624	−0.3532
	0.9827	0.0576	0.0173

These results demonstrate the well known instability of the Leontief dynamic model. In Livesey (1976) it was shown how the state space representation of the model preserved the properties of the matrix $B(I-A)^{-1}$. We are particularly interested in the one root which has a non-negative eigenvector since this is the upper bound on the economy's growth rate. The values given above imply that the maximal growth rate for the UK economy in 1972 was 28 per cent. Not surprisingly this accords with Barker's results, for the same data, used at a greater level of disaggregation. The level of disaggregation may not be important since in Livesey (1976) it is suggested that the order of the minimal realisation for the Leontief model is equal to the number of investment assets distinguished, which in Barker (1971) was three.

Although the empirical magnitude of the maximal growth rate is of some interest, it is purely academic since it is associated with a zero growth rate for the consumption level. We are much more interested in situations where the rate of growth of consumption is strictly positive, and in observing the relationship between it and the rate of growth of output. This is a problem ideally suited for feedback theory.

We need to fix a proportional relationship between gross output and final demand. It seems sensible as a first approximation to assume that the 1972 relationship, given in Cambridge, Department of Applied Economics (1970: Table V.1), remains constant. In other words,

$$f_t \propto \underline{K} q_t = \begin{bmatrix} 0.2277 & 0 & 0 & 0 \\ 0 & 0.0972 & 0 & 0 \\ 0 & 0 & 0.6578 & 0 \\ 0 & 0 & 0 & 0.6063 \end{bmatrix} q_t. \quad (2.13)$$

This is a diagonal matrix and is clearly only an approximation to the exact classification converter but it is satisfactory for this exercise. If in equation (2.11) we set $f_t = gKq_t$ then we are applying positive feedback with a gain of g. As g increases so the level of consumption increases and we would expect the rate of growth of q_t, gross output, to decline. When $g = 0$ there is no consumption and when $g = 1$ the level of consumption, for gross output equal to its 1972 level, is also equal to its 1972 level. Using the estimated four-sector Leontief model given above, we obtained:

% growth rate of gross output	*feedback gain*	*% gross savings rate (i.e. including depreciation)*
27.9	0	100
25.3	0.2	85
22.5	0.4	70
19.2	0.6	56
15.0	0.8	41
11.9	0.9	34
10.2	0.94	31
8.1	0.98	28
6.7	1.0	26
3.1	1.04	23
0.6	1.06	22
−2.9	1.08	20

With a feedback gain of one we are nearest to the normal allocation of resources between consumption and investment. It is therefore interesting to know the relative magnitude of the capital assets given by the eigenvector associated with the growth rate of 6.7 per cent and a savings rate of 22 per cent.

Relative Magnitude of Capital Stocks

estimate of 1972 actual value 22 % savings rate	*with savings rate of 100 %*	*with savings rate of 26 %*
1.0	1.0	1.0
0.56	0.53	0.54
0.11	0.12	0.12

In other words the allocation of capital between assets is not too different from that associated with faster rates of growth. Even with a feedback gain of one the Leontief model remains very unstable. The two other eigenvalues are -3458 and 3.4.

One way to stabilise the Leontief model is to design a feedback control matrix which will place all the eigenvalues within the unit circle. Using an algorithm due to Basil Kouvaritakis one can construct the matrix,

$$\begin{bmatrix} 27.1 & 21.9 & -335.2 \\ 22.0 & 17.8 & -271.7 \\ -333.1 & -269.6 & 4121 \\ -249.9 & -202.3 & 3092 \end{bmatrix} \tag{2.14}$$

which, when used to generate final demand as a function of the capital assets, yields a system with eigenvalues -0.53, 0.24 and 1.37. We have not completely escaped from the problems of instability since the eigenvalues are critically dependent upon the accurate implementation of the gain matrix (2.14). Any error, even in the fourth significant figure, will result in unstable eigenvalues.

Feedback theory allows us to explore the trade-off between the rate of growth and the level of consumption. It also underlines the inescapable instability of the Leontief system.

III NONLINEAR MODELS—SOME TENTATIVE CONCLUSIONS

The feedback theory, which we have used in the previous two sections, is only directly relevant to linear models. This is of course also true of many of the main results in optimal control theory. There is no reason why feedback theory cannot be applied to nonlinear models using local linearisation. This is the technique which has been successfully employed in optimal control theory applications. In essence the approach would be to generate a locally valid linearisation of a nonlinear model and to design a stable feedback controller for this representation.

It is tempting to conclude that this feedback theory approach gives us so little when compared to the scope of problems which can be tackled using optimal control theory. In particular one has in mind the ability of optimal control theory to deal with stochastic problems. Uncertainties are undoubtedly a crucial feature of economic applications, and many people believe that these can only be properly dealt with using stochastic control theory. Feedback theory only appears to tell us whether a model is locally stable. We may be tempted to feel that this is not a significant addition to our knowledge. However, when we are faced with disturbance processes the properties of which we do not fully understand, we have to rely entirely upon the characteristic modes of the system to drive it back towards equilibrium.

Consider for a moment a horse-racing analogy. The use of stochastic control theory has the potential danger that we might be breeding a very highly strung racehorse. On a fine day with a firm course this horse will perform admirably and will outstrip the field. In economics however we are very uncertain about the course over which our horse will have to race. From time to time the horse will be subjected to considerable disturbances and may indeed fall at one or two fences. It is not unknown for racehorses to have to be destroyed following accidents on the course. Economic policy derived using feedback theory may produce a large and plodding carthorse. However, this carthorse is much more likely to be able to withstand shocks and even to plough through one or two fences which suddenly rear up as obstacles on his course.

Stochastic control theory is only optimal when we have a very narrowly defined class of problems. Most economic applications do not fall into that category and so the stochastic control results produced for most economic models are not optimal. We cannot be certain how these economic policies derived using stochastic control theory will perform under all situations. Although it is clear that feedback theory will not revolutionise the present line of development being pursued by economists using optimal control theory, we feel that it will nevertheless have some useful insights to offer of a very cautionary nature.

REFERENCES

Aoki, M. (1973), 'On Sufficient Conditions for Optimal Stabilization Policies', *Rev. Econ. Stud.* vol. 40, pp. 131–8

Barker, T. S. (1971), 'A Maximum Sustainable Growth Rate for British Industrial Outputs', *Rev. Econ. Studies*, vol. 38, pp. 369–76.

Livesey, D. A. (1976), 'A Minimal Realization of the Leontief Dynamic Input-Output Model', in Polenske, K., and Skolka, J. *Advances in Input-Output Analysis* (Ballinger Books, Mass.) pp. 527–41.

MacFarlane, A. C. J. (1975), 'Relationships between Recent Developments in Linear Control Theory and Classical Design Techniques, *Measurement and Control*, vol. 8, pp. 179–87, 219–23, 278–84, 319–24 and 371–5.

Phillips, A. W. (1954), 'Stabilisation Policy in a Closed Economy', *Econ. J.*, vol. 64, pp. 290–323.

Cambridge University Department of Applied Economics (1970), 'Exploring 1972 with Special Reference to the Balance of Payments', no. 9 in *A Programme for Growth* (Chapman & Hall, London).

5 Control Methods and Quantitative Economic Policy

Mark Salmon and Peter Young

The word optimal carries with it the suggestion that the system has desirable properties in general, but of course this need not be the case. In particular, [it can be shown] that the optimal system has properties which cannot usually be achieved in practice and may have properties that are highly undesirable.

H. H. Rosenbrock and P. D. McMorran, 'Good, Bad or Optimal', *IEEE Trans. Aut. Control*, vol. AC. 16, no. 6, 1971.

1. INTRODUCTION

Our intention in the present paper is to emphasise certain issues which have been neglected in the literature but which will have important implications on the *practical* application of control methodology to economic management.

The literature relating control theory to problems of applied macroeconomic policy-making may be separated into two indistinct areas. The first and most popular is concerned with the exploration of an econometric model's properties in terms of the implied effectiveness of alternative policies. The second, much less popular but possibly more important, topic is concerned with the problems of forming *robust* economic policy for the *real* world.

The difference between the two approaches lies mainly in the extent to which note is taken of the possible structure of the *actual* system of the real world in contrast to the particular *model*, as perceived and used by the analyst. Whilst examination of a model's properties within the optimal framework gains its strength through the evaluation of the 'best' operation of alternative policy instruments, little or no research has been directed towards assessing the *sensitivity* of these apparently optimal policies to 'model misspecification'. Too often, the alternative policies are generated simply by considering changes in the weighting of variables in the welfare function. Yet it is the examination of the sensitivity that is most relevant to the second kind of control study; namely the generation of robust policy strategies in the real world. For instance, it is quite possible that while fiscal measures may be better, in terms of welfare loss, for achieving a particular target, a near equivalent monetary policy may be less sensitive to model misspecification and thus a safer bet.

In the present paper we will argue that it is impossible to separate meaningfully optimisation studies based on models of an economic system and the application of the results of these studies to real-world problems. Optimal solutions based in isolation on the dynamic properties of a model will mean little if, as in the economic case, the system is badly defined and the model open to misspecification. If these optimal techniques are to be taken seriously in this situation and considered as an aid to policy formation, then they must take note of uncertainty and possible model misspecification. In particular control strategy designs should be developed that are, to some degree, 'robust' in the face of these problems.

The practical success of feedback control in physical and engineering systems can be attributed largely to the interplay between theory and practice: where this interplay has not been given adequate attention, the practical exploitation of new theoretical concepts tends to be retarded. Thus it is largely a myth that the many developments in optimal control theory have had a large impact on practical control system design. Indeed many economists may be unaware that, even in the case of engineering control problems where systems are relatively well defined and usually man-made, optimal control developments have so far had very limited success in terms of proven practical control system design. This lack of success has resulted almost entirely from the 'gap' which developed over the nineteen-sixties between the more esoteric developments in optimal control theory and the practical implementation of control strategies in the real world. Whilst the optimal Linear Quadratic Gaussian (LQG) approach, for example, provides a neat theoretical framework pleasing to the applied mathematician, it tends to obscure certain practical aspects of the control problem unless applied with great care (see e.g. Rosenbrock and McMorran, 1971).

In order to clarify our presentation and avoid possible confusion, we will largely restrict any mathematical elaboration of the above points to the deterministic control situation. Thus, while we will refer to the problems introduced by uncertainty and model misspecification we will illustrate these discussions in mainly deterministic terms. It should be emphasised, however, that all of our suggestions can be incorporated in the stochastic case. It is also important to note that it is the distinction between uncertainty in the stochastic case and robustness to misspecification that is our prime concern. In the stochastic case, the additional information built into the model through probability specifications will be of considerable value, but it is unlikely that a subsequent control system designed using current optimal methodologies would be fully satisfactory in a *practical* sense if model misspecification results from the stochastic uncertainty, as it almost certainly will. We argue that it is this latter *consequence* of uncertainty which is most relevant to the economic policy maker in practical terms.

Before considering the problem of model misspecification, we discuss differences in the concept of the appropriate model that have arisen in the control and econometric literature. We suggest that the econometrician concerned with economic control should be concerned with 'modelling for control', even though his background experience may have led him towards models that have a primarily forecasting role. We then consider the basic operations of any multivariable control method and the very specific, but relatively minor, place of the optimal LQG approach in this context. This leads naturally to a discussion of the limitations inherent in current optimal state space analysis for economic systems and some preliminary suggestions for obviating these limitations. Finally, we consider a simple example through which we hope to make more apparent some of our preceding remarks.

2. MODELLING FOR CONTROL

2.1 The Concept and Nature of a Control Model

Any model that is to be used for operational control of the economy must be first and foremost a '*control model*'; a model which attempts to specify as clearly as possible, within the limitations imposed by the badly defined nature of the economic system, those mechanisms which appear to be operative between the *control inputs* or instruments and the *control variables* that they are meant to control. Although most large-scale econometric models are aimed at producing a forecasting tool and are not specifically designed for control, it may well be argued that a forecasting model would automatically be a good control model and vice versa. This is not necessarily the case: a forecasting model may contain much information on the mechanisms operating between the control instruments and the control variables but such information is not essential. For example, a simple time series model can often provide adequate forecasts of economic variables, but it may only be of limited use in control design. Similarly the large econometric forecasting model may have within it much information of potential use for control design but this information may not be readily available in a form useful for control analysis. Pierce [18] has emphasised these points in considering the utility of a version of the FMP model of the US economy in this context.

In order to consider the requirements of the control model, let us start from an assumption that the economic mechanisms can reasonably be modelled by a set of nonlinear state space differential equations, [19] written as follows:

$$\underline{\dot{x}} = f(\underline{x}, \underline{u}, \underline{d}, \underline{\xi}, t) \tag{1}$$

where

- $\underline{x}$ is the state vector
- $\underline{u}$ is the control input vector, composed of elements which can be considered as control instruments,
- $\underline{d}$ is a disturbance vector representing those other exogenous disturbances affecting the system but not being considered as control variables, and
- $\underline{\xi}$ is an r dimensional vector of stochastic disturbances affecting the system.

Let us further suppose that the vector of economic variables that are to be controlled, y, is related to the state vector $\underline{x}$ by another non-linear relationship, the observation equation, of the form:

$$\underline{y} = \underline{g}(\underline{x}, \underline{\zeta}, \underline{t}) \tag{2}$$

where $\underline{\zeta}$ is a p dimensional vector of stochastic variables affecting this output relationship, possibly due to measurement uncertainty.

In control analysis, it would be unusual to work directly with equation (1) and (2) since there is no unified theory of control for such nonlinear systems and, in any case, no evidence in practical control systems design that such a complex formulation is normally required. In general, two approaches are favoured; both depending on the generation of a 'linearised control model'. In the first and academically most attractive approach, a nominal solution of equation (1) is obtained by solving for equilibrium, having chosen that equilibrium concept (amongst possibly many) which seems most suitable to the control problem. The system and observation equations are then linearised about this solution (which could be static but might well be dynamic) to yield a 'linearised' small perturbation model as follows

$$\begin{aligned} \delta \underline{x} &= F\delta \underline{x} + G\delta \underline{u} + E\delta \underline{d} + \delta \underline{\xi} \\ \delta \underline{y} &= H\delta \underline{x} + \delta \underline{\zeta} \end{aligned} \tag{3}$$

where E, F, G and H are appropriately dimensioned *and possibly time-variable* matrices, while the prefix delta is used to imply small perturbation values of the variables about the nominal solution.

In the second, more usual and pragmatic approach, a set of equations such as (3) is assumed to apply and the nature of the nominal solution is usually implied by the nature of the observed variables. For example, in an economic context, the analyst using this second approach might well consider equations such as these to apply for the *changes* of economic variables about the long-term (low-frequency) trends observed in the economic data. In other words, he would decompose the system into long-term low-frequency behaviour and short-term small-perturbation changes around the long term.

Since the economic data are available only in sampled form, it is usual to deal not with equations (3) but with discrete time equivalents of these equations of the general form

$$\underline{x}_{t+1} = A\underline{x}_t + B\underline{u}_t + D\underline{d}_t + \underline{\xi}_t \tag{4}$$

$$\underline{y}_t = C\underline{x}_t + \underline{\zeta}_t \tag{5}$$

where A, B, C and D are appropriate matrices, again with possibly time-variable coefficients. Here the prefix delta has been omitted both for convenience and because this is usual in most analysis. (This omission of the prefix can be dangerous, of course, since it may make the analyst lose sight of the small-perturbation nature of the equations – see section 4.)

Referring to (4) and (5) we see that the relationship between the control input vector $\underline{u}_t$ and the output vector $\underline{y}_t$ can be obtained by applying for z or forward shift operator, i.e. $z\underline{x}_t = \underline{x}_{t+1}$, to these equations, to yield the following

$$\underline{x}_t = [zI - A]^{-1}B\underline{u}_t + [zI - A]^{-1}D\underline{d}_t + [zI - A]^{-1}\underline{\xi}_t \tag{6}$$

$$\underline{y}_t = C\underline{x}_t + \underline{\zeta}_t \tag{7}$$

Substituting from (6) into (7) we obtain

$$\underline{y}_t = C[zI - A]^{-1}B\underline{u}_t + C[zI - A]^{-1}D\underline{d}_t + (C[zI - A]^{-1}\underline{\xi}_t + \underline{\zeta}_t) \quad (8)$$

where $C[zI - A]^{-1}B$ is the multivariable control input-output transfer function of the system which can be written alternatively as

$$C[zI - A]^{-1}B = C\frac{\text{Adj.}\,(zI - A)}{\det\,(zI - A)}B = \frac{1}{D_1(z)}N(z) = G_1(z) \quad (9)$$

Note that equation (8) can be written more simply as

$$\underline{y}_t = \underline{\hat{y}}_t + \underline{d}^*_t + \underline{\eta}_t \quad (10)$$

where $\underline{\hat{y}}_t$ can be considered as that contribution to the output vector arising from $\underline{u}_t$, i.e.

$$\underline{\hat{y}}_t = C[zI - A]^{-1}B\underline{u}_t = G_1(z)\underline{u}_t \quad (11)$$

while

$$\underline{d}^*_t = C[zI - A]^{-1}D\underline{d}_t \quad (12)$$

and

$$\underline{\eta}_t = (C[zI - A]^{-1}\underline{\xi}_t + \underline{\zeta}_t) \quad (13)$$

Equation (10) provides a model for the output vector which, due to the linearity of the system, is seen to consist of three terms; the first is indicative of that part of $\underline{y}_t$ which can be directly attributed to the control variable $\underline{u}_t$; the second quantifies the effect on the output variables of those other, but uncontrolled, exogenous variables $\underline{d}_t$ affecting the system; and finally, the third term provides a description of the manner in which the output variables are perturbed by the various stochastic inputs disturbing the system. All three terms in equation (10) are, of course, important to the control analysis, but it is the first term which is of primary importance because it helps to define the mechanisms operating between the *controlled input* variables and the *controlled output* variables. Thus while he will wish, if at all possible, to quantify in some manner the transfer functions characterising all three terms on the right-hand side of equation (8), the control analyst will give first priority to the identification and estimation of a model which describes the mechanisms implicit in the first term. In this sense, we might refer to equation (11) as a 'control model' of the system.

To contrast this concept of a control model with a more conventional econometric concept of a 'forecasting model' we can see that the accurate prediction of $\underline{y}_t$ is more likely to require a good knowledge of all three contributing terms on the right-hand side of equation (10) since such prediction implies the forecasting of all changes in $\underline{y}_t$ not just those due to $\underline{u}_t$. Thus, where the control analyst may well be able to design a *reasonable* control system given only a knowledge of the first 'control term' it would be surprising if the econometric model based only on this first term would yield acceptable forecasting results.

It should be emphasised here that we are *not* saying that the control analyst *only* requires the first control term in equation (10); merely we are indicating that he might well be able to *manage* with only this degree of information. Clearly a knowledge of other factors affecting the system, as defined in the present context by the second two terms of equation (10), will help him to improve his control system design; and this is most important. But the control term provides *the most significant information in control system design terms* and so it is this aspect of the overall model that the control analyst has to emphasise initially. Thus he might well argue that the *robust* estimation and identification of this part of the model is of greatest priority; while the econometric forecaster, with his different aims in mind, would tend to concentrate more on the *statistically efficient estimation* and prediction of y_t.

Most economic control studies tend to recognise, if only by force of analytical necessity, that a model to be used in optimisation and control calculations must have rather different characteristics to those of the larger forecasting model.[1] But, as we have seen, the differences between a control model and models intended for other uses, may derive from different considerations than those of analytical convenience.

Certainly from this point of view it is clear that the control model need not be large; indeed the idea of a possible 'Law of Large Systems' together with the associated concept of 'dominant mode' dynamic behaviour [35] would suggest that the control model could be quite modest in relation to most practical forecasting models. (See also concluding comments in [18].)

2.2 The Identification and Estimation of Control Models

The implicit recognition (through practical experience) of the differences in concept and nature between models suitable for control and forecasting has led control engineers in the physical sciences to follow different approaches to the problems of identification and estimation. We briefly mention four major areas where these differences arise, whilst accepting that the physical systems approach may not be totally acceptable to the problems faced by the econometrician.

(i) The Use of Small Perturbation Linearised Models Since the control analyst tends to use the small perturbation linearised approach he also tends to work with 'small perturbation data'. Thus, when confronted with variables which, like many macroeconomic variables vary over a large range of values with clearly defined trends, he will feel the need to remove these trends (either by differencing or preferably high pass filtering or through the use of a separate long-term model (see section 4)) prior to *linear* modelling. Note that this transformation is made so that an appropriate model of the system is analysed without the possibly confounding effects of the long-run and potentially nonlinear characteristics of the true data generation process.

The econometrician will sometimes adopt similar 'stationarity' transformations but they are usually chosen solely on statistical grounds. However as long as short run disequilibrium economic theory is unable to provide sufficient *a priori* justification for the kind of dynamic separability imposed by engineers, it appears that many modellers will be content to analyse the data 'as is' and yet will often still adopt a linear model. The problem is, of course, exacerbated in estimation by an increased noise to signal ratio which can be induced by such "stationarity" transformations when applied to economic data that often is already relatively uncertain. More sophisticated high or band pass filters do give considerably more flexibility in this respect to minimise the high frequency noise amplification problem. We note however, that the common stationarity transformations such as differencing or least squares trend removal are not uniform over all frequencies. As a result the question of the relevance of dynamic separability remains and requires substantially more theoretical analysis.

(ii) Differing Concepts of Identifiability The concept of structural identification in econometrics can be considered to develop basically from a relationship between the probability density or likelihood function associated with the model, a set of identifying restrictions and their unique solution. In the control literature, the concept of identifiability may be thought of as being concerned with whether, on the basis of a *given data set*, it is possible to estimate a unique set of parameters associated with the specified model without ambiguity. Following from the ability to perform planned experiments directly on the system, the control analyst in physical systems has tended to be concerned not solely with the structure of the model but also with factors such as the nature of the input perturbation signals and the overall signal topology within a feedback context. In particular the requirement of a 'persistently

exciting' input so that all elements of the system are sufficiently stimulated becomes one condition of 'identifiability' [14] [31]. Other conditions are concerned with the controllability together with observability of the system and the invertibility of the noise process. Put simply, these conditions are analogous in regression terms to the conditions on the rank and the non-singularity of the regressor cross-product matrix. The wider implications of these conditions of identifiability and their relationship to consistent estimability have been discussed, for example, in [6].

(iii) Differences in Estimation Methodology Here again, there are many differences in approach to model estimation although there has been a tendency in the last few years for both control analysts and econometricians to better appreciate each other's point of view. In general, the differences in approach derive directly from the differences in the requirements of control and econometric models as well as through a recognition of the characteristics of the available data. Thus the control analyst tends to utilise methods which emphasise the robust estimation of the control model characteristics while the econometrician has emphasised methods which are statistically efficient.

By the nature of their applications control engineers have been forced to consider the problem of noise or uncertainty on the observed system variables. Thus they have been naturally directed towards an errors-in-variables model formulation and techniques more suitable to this problem both in the observation space [29] and the state-space (via state-parameter estimation) [3]. The formulation of the model in terms of state equation and an observation equation allows more direct examination of how noise in all its forms affects the system than is achieved in the standard econometric errors in equation model formulation.

(iv) Recursive versus Block data Processing While most econometric model estimation is based on block data processing, the concept of recursive estimation (that is, the analysis of the time series data, one observation at a time – see e.g. [29], [23]) is comparatively little known, even though it has had an important influence on estimation methodology in the control literature. We could discuss this point in great detail but it will suffice to say that, once again, the concept of 'modelling for control' has appeared to have an important influence on this approach to estimation. From the conceptual small perturbation analysis, where the model matrices are seen to have possibly time-varying coefficients, to the

idea of adaptive control, where the control system is adapted to changes in the system dynamic characteristics, the control analyst tends to view the system model as inherently time-varying. And since recursive estimation is in an ideal form for the estimation of time-variable parameters, it has obvious attractions for him. In addition, recent developments in recursive estimation methodology have yielded asymptotically efficient algorithms for single and multivariable systems [36] [37].

3. APPROACHES TO CONTROL DESIGN

The literature in the control of economic systems has been so completely dominated by the optimal LQG case that the unwary economist would be justified in believing that this is the only approach to control design used by control engineers that has any relevance to economic problems. Since this is certainly not the case, there exists a definite need to clarify the situation and indicate the alternatives and how they are related. Much of what we shall discuss below reflects and generalises the basic ideas behind control design discussed by Phillips [16], [17]. We feel that the implications of some of the points raised by Phillips have been lost in the more recent developments.

Whilst the optimal LQG case may provide a most elegant conceptual framework for the economic problem, it is well known – if again not often discussed – that the resulting optimal control strategy may not be a good, nor even satisfactory strategy in general terms (that is, of course, in any other sense apart from the narrow definition of the minimisation of the criterion function). This is a separate matter from the practical difficulties of implementing the control in the political and technical environment. It is more a matter of the inherent qualities of the controlled economy in terms of its robustness, developing from sensitivity to changes in structure or misspecified criterion function in policy construction. Chow in his recent book [4] mentions this point in passing, but devotes little or no time to the analysis of the question.[2] This is partially because the optimal LQG approach subsumes to a degree the important question of control design and control system structure. In accepting the convenience of the optimal framework, a range of important questions relating to the characteristics of the control have, in effect, been ignored. In less systematic control procedures, these questions have to be faced directly as a principal part of the control design process.

This accords to some degree with the scepticism of the economist in government that such a rigid approach to his problems is not feasible. The design of a suitable budget requires some degree of finesse and substantial manipulation prior to the final control selection; and this is exactly what a practical control engineer in industry would say with regard to his problem. It is a misconception to think that any control method can obviate these important practical issues. The fact that this impression is prevalent is a natural consequence of the uniqueness of optimality. Given the desired trajectories, criterion function, weights, and the econometric model, the control algorithm will provide one and only one optimal solution. What the policy-maker may prefer to see, in addition to a range of optimal solutions for different criterion function weights, is a range of feasible alternative controls in response to varying degrees and types of misspecification so that he may assess the sensitivity and robustness of alternative policies in this respect.

In the larger model situation, the analysis of these questions would of course, be more complex, requiring the careful identification of the principal control links in any particular case. Having isolated these relationships, the *sensitivity* with respect to variations in the critical parameters (for instance) could be seen by altering their values in some systematic manner. This would then provide a means for assessing the importance of an implied misspecification by observing the resulting behaviour of the controlled model under the assumption that the control law remains as generated by the original parameters. The analysis would not stop at this stage, of course; rather the control would be redesigned to be as robust as possible in the face of the misspecifications determined to be most important. The *existing* optimal approach provides no information towards this end.

We take up this point below but first consider questions relating to the appropriate framework for control design, in particular noting that the optimal approach is only *one* way of formally generating a control strategy. The following subsection then summarises the general theory by which any state-variable feedback control method achieves its results, and then we suggest how this may be extended to incorporate the design of robust controllers.

3.1 Multivariable Control: an Overview

Consider an economic system where *small perturbation* behaviour can be described by equations (4) and (5). For clarity of exposition we shall ignore stochastic disturbances and non-control exogenous variables. We then see that the relationship between the output vector $\underline{y}_t$ and the control variables or instruments $\underline{u}_t$ is given by equations (8) and (11) as

$$\underline{y}_t = C[I - A]^{-1} B\underline{u}_t = G_1(z)\underline{u}_t \qquad (14)$$

where $G_1(z)$ is the pxm multivariable transfer function matrix (MTF) whose elements are seen from equation (9) to be rational algebraic functions of z, i.e. ratios of polynomials in z. These ratios of polynomials or transfer functions have a *common* denominator $D_1(z) = det(zI - A)$ which is the characteristic polynomial of the system. From the figure below we can see clearly that the ith output, $i = 1, 2, \ldots p$, is, in general, affected by all of the m control variables via the transfer functions $\frac{N_{ii}(z)}{D_1(z)}$, $j = 1, 2, \ldots m$. We note that the dynamic behaviour of the system will be strongly dependent on the nature of the characteristic polynomial $D_1(z)$;

$$\begin{bmatrix} y_1 \\ y_2 \\ \cdot \\ \cdot \\ \cdot \\ \cdot \\ \cdot \\ \cdot \\ \cdot \\ \cdot \\ y_p \end{bmatrix} = \frac{1}{D_1(z)} \begin{bmatrix} N_{11}(z) & N_{12}(z) \ldots & N_{1m}(z) \\ N_{21}(z) & N_{22}(z) \ldots & \ldots \\ \cdot & \cdot & \cdot \\ \cdot & \cdot & \cdot \\ \cdot & \cdot & \cdot \\ \cdot & \cdot & \cdot \\ \cdot & \cdot & \cdot \\ \cdot & \cdot & \cdot \\ \cdot & \cdot & \cdot \\ \cdot & \cdot & \cdot \\ N_{p1}(z) & \ldots\ldots & N_{pm}(z) \end{bmatrix} \begin{bmatrix} u_1 \\ u_2 \\ \cdot \\ \cdot \\ \cdot \\ \cdot \\ \cdot \\ \cdot \\ \cdot \\ \cdot \\ u_m \end{bmatrix}$$

in particular on the eigenvalues of the characteristic equation $D_1(z) = 0$, which are often referred to as the poles, whose positions in the complex plane determine the natural modes of dynamic behaviour of the system. The relative importance or 'dominance' of these modes on the behaviour of the input variables is then determined by the nature of the numerator polynomials, $N_{ij}(z)$, and the resultant residues associated with each mode [24]. Usually only a small number of the natural modes will have an appreciable effect on the output behaviour of the system; in other words although nominally of high order n, the system may tend to behave dynamically as though it were of much lower order $n_1 < n$. This concept of modal dominance is clearly important at the estimation and identification stage as well as in the formulation of control design [35].

The optimal LQG formulation of the control problem leads mathematically to a control law in which the values of the control variables are determined as linear functions of the state variables, i.e.

$$u_t = -K_t \underline{x}_t + \underline{g}_t \tag{15}$$

where K_t is an mxn state variable feedback gain matrix and $\underline{g}_t$ is a function of the desired trajectories specified in the cost function. (Notice that although the elements of K may well be time-varying they will be frequency independent, i.e. independent of z.) The effect of such state variable feedback is seen by substituting from equation (15) into (4) and then obtaining the MTF between $\underline{y}_t$ and $\underline{g}_t$ as

$$y_t = \frac{1}{D_2(z)} M_2(z) \underline{g}_t = G_2(z) \underline{g}_t \tag{16}$$

where $D_2(z) = det\ (zI - A + BK)$ is the 'closed loop characteristic polynomial'. By direct analogy with the uncontrolled 'open loop' case we now see that the eigenvalues of the closed loop characteristic equation $D_2(z) = 0$ define the poles of the closed loop system and that the principal dynamic effect of the control law is to move the open loop poles to these new positions in the complex plane. The degree to which the open loop system dynamics can be changed in this manner is demonstrated by the following fundamental theorem (see [27]).
Assume that A and B are real, and let $\Lambda_n = [\lambda_1 \ldots \lambda_n]$ be an arbitrary set of n complex numbers (desired closed loop poles or eigenvalues), then

Theorem
If the pair (A, B) is completely controllable, in the sense that the controllability matrix $[B \vdots AB \vdots \ldots \vdots A^{m-1}B]$ has rank n, then there exists a real matrix K such that the roots of det $(zI - A + BK)$ are the set Λ_n.

In other words, this theorem tells us in the present context that if the system is controllable, it is possible to choose a feedback gain matrix K which, in theory, will place the closed loop poles at arbitrarily specified positions in the complex plane. In practice this result is only meaningful provided the system remains linear under closed loop control. But within these limitations the theorem indicates the feasibility of constructing procedures for closed loop pole assignment. Numerous such procedures have been developed and the reader can find an introduction to the literature on the subject in a number of references (e. g. as listed in [15], [20].)

It is worth noting here that in the optimal LQG case, the control strategy is determined without direct reference to the full implications of the positions assumed by the closed loop eigenvalues, but with more direct concern to questions of the appropriate economic trade-offs through concentration on the effects that the policy has on the value of the welfare function. This is one reason why unsatisfactory control strategies may result from the optimal LQG approach if due care is not taken to analyse the dynamic properties of the controlled system.

Further, this algebraic interpretation of the optimal policy is clearly important in dynamic terms and provides additional insight into the nature of optimal policy in closed loop feedback control terms. Indeed the algebraic results are, in a practical sense, more general: state-variable feedback control is seen to be not only important because it happens to coincide with the optimal LQG policy but also because of the considerable measure of control it provides over the closed loop pole locations. It is not surprising, therefore, that the control literature on state variable feedback (SVF) is not limited to optimal interpretations but has been concerned with other, nominally 'non-optimal', approaches to control system design including

(*a*) those algebraic methods concerned with the assignment of the closed loop eigenvalues (poles) to specified positions in the complex plane – the so called 'pole assignment' or 'modal control' methods [20];
(*b*) the achievement of exact model matching [26]; or
(*c*) decoupling of multivariable systems [9].

Reference to this more extensive literature in SVF reveals certain limitations of SVF, at least as it has been applied until very recently. There has always been some measure of dissatisfaction with SVF in the control engineering world, but it was not until the beginning of the 1970s that the discontent emerged in the control literature [15]. These criticisms culminated recently in a paper by Horowitz and Shaked [10], who deliver a strong attack on the SVF approach to control system design which they compare unfavourably with the alternative 'transfer function methods' of conventional control theory.

But Horowitz and Shaked are themselves prone to rather dogmatic statements and have a tendency to overemphasise their case. Whilst SVF certainly has its disadvantages it also has important implications for control system design which cannot be ignored. In particular the great degree of control over the closed loop dynamics imparted by state-variable feedback has important implications on control system design and should always be taken into account. Further, the relative ease with which multiple objective problems may be synthesised through the cost function should not be overlooked.

What seems to be required is a general approach to control analysis and design which combines the best aspects of both the conventional (frequency response) and modern (state space) control methodologies in a unified framework; a framework which attacks the practical problems of multivariable feedback control in the presence of uncertainty. While a satisfactory unified approach to multivariable control of this type is certainly not available at this point in time, there are existing design procedures which seem to offer some hope that such an approach will develop in the future.

3.2 Design for Robust Economic Control

It is true of course that much useful work has been carried out in recent years particularly addressing the problem of economic control in the face of uncertainty. But in this work, the examination of the sensitivity of the implemented control to error between the assumed model and the actual system has been neglected. Thus, while Chow [4] discusses the problem of 'multi-period decision under uncertainty', he rarely evaluates quantitatively the effects of model misspecification which may well follow from this uncertainty.

As can be noted from the definition of the state, SVF with constant or time-variable gains in the SVF gain matrix implies 'proportional plus derivative' control. Unless applied to problems formulated in stochastic terms with certain non-stationary disturbances, however, basic SVF control approach does not yield a control law which utilises direct 'integral action', i.e. the control law contains no terms which are bound directly to the integral of error between the desired target levels for the system variables and the *actual* levels attained.

It can be argued of course that the integral of error between the desired target levels and the *model* variables is inherent in the intercept term g_t. But this is clearly of little practical utility in the presence of model misspecification since minimisation of the model-target error will almost certainly imply the existence of an error or 'offset' between the *actual* outcomes and their specified targets if the model is not an *exact* representation of the actual system.

There are a number of methods available for overcoming the problem of offset in multivariable SVF control systems. Essentially this problem is one of removing offset errors for general classes of input and target specification. Perhaps the simplest approach is the multivariable equivalent of the single variable PID method (Proportional, Integral and Derivative Control).

Integral of error action can be introduced through the augmentation of the state space to include a vector of additional state variables $\underline{r}_t$ defined as the integral of error between the output vector $\underline{y}_t$ and the specified target levels for this vector $\underline{y}_r$ where the 'integral error' vector $\underline{z}_t$ is defined as

$$\underline{z}_{t+1} = \underline{z}_t + (\underline{y}_r - \underline{y}_{t+1}) \tag{17}$$

so that the augmented system becomes

$$\begin{bmatrix} \underline{x}_{t+1} \\ \underline{z}_{t+1} \end{bmatrix} = \begin{bmatrix} A & O \\ -CA & I \end{bmatrix} \begin{bmatrix} \underline{x}_t \\ \underline{z}_t \end{bmatrix} + \begin{bmatrix} B \\ CB \end{bmatrix} \underline{u}_t + \begin{bmatrix} O \\ I \end{bmatrix} \underline{y}_r \tag{18}$$

Provided this augmented system is controllable, then from Theorem 1 (section 3.1), it is always possible to define a control law

$$\underline{u}_t = -[K_{pd} \vdots K_i] \begin{bmatrix} \underline{x}_t \\ \underline{z}_t \end{bmatrix} = -K_{pd}\underline{x}_t - K_i\underline{z}_t \tag{19}$$

where K_{pd} and K_i are respectively, the 'proportional and derivative', and 'integral' feedback gains which will locate the eigenvalues of the closed loop system at specified positions in the complex plane. And if K_{pd} and K_i are chosen to ensure that the eigenvalues are within the unit circle, so that the system is stable, then a steady state will exist. In the steady state $\underline{z}_{t+1} = \underline{z}_t$ and, from equation (17) we see that $\underline{y}_t = \underline{y}_r$ and the system will have attained the desired target levels. It will be noted that, provided the inevitable error between the assumed model and the actual system does not lead to closed loop instability, then the target levels will *always* be achieved in the steady state; in other words, we might say that the closed loop steady state is *invariant to model misspecification.* Of course, economic control with the problem of the resolution of multiple objectives we would need to modify the severity of this type of control action.

In this regard it is easy to consider this multivariable PID system within an optimal context [30] and, in this manner, differential weighting can be applied in the cost function both to the basic state variables and the integral of error states. This would seem to allow for the kind of compromise in achieving target levels that is necessary in economic systems.

This latter idea has been taken to its logical conclusion by Johnson, who has considered not only the problem of achieving general target specifications [11] but also the related problem of maintaining good target 'tracking' abilities in the face of general types of input disturbances [12].

Using a similar method of state vector augmentation to that used for improving target following capability, he is able to obtain control system design which tends to counteract the effect of fairly general classes of disturbing signals and may even exploit the effects of these disturbances to achieve better closed loop regulation about the specified target paths. The parallel between Johnson's approach to disturbance control and the optimal LQG problem can be drawn, of course, but by *concentrating* attention on the need for specific considerations of target specification and external disturbances, Johnson yields a design procedure with greater physical appeal and flexibility. And unlike the LQG approach, there is certainly little danger of forgetting the importance of these factors in control system design.

It is worth noting here that Fuller [8], in considering the optimum control of single input-single output systems in the face of low frequency stochastic disturbances, has found by frequency domain analysis that, if observational (i.e. measurement) error and constraints on the control action are not considered adequately, then the resulting optimal control system has very poor stability margins. He shows, however, that by taking these factors into account in a specific manner the optimal system is much improved and approximates a 'conventional' type such as proportional plus integral. Fuller points out that his results yield similar designs to those obtained by incorporating integral action in the controller at the outset of design, although he contends that his approach is more theoretically satisfactory. It could well be that an extension of his approach to the general multivariable case would have some appeal for economic control theorists.

All of the more recent developments in SVF control discussed above depend for their success on increasing the order of the system by the augmentation of the state vector to include additional state variables. This is equivalent in conventional control terms to the use of dynamic compensation filters (i.e. transfer functions in z) to improve the overall frequency response of the system and they tend to combat the criticism that SVF concentrates over much on the rather inflexible use of constant, in the frequency independent sense (i.e. K independent of z), compensating elements. The SVF advocate would argue that the alternative approach using output feedback through an observer, or in the stochastic case a Kalman Filter, is all that is required. It is clear that such an approach can certainly provide good guidance to control system design and should figure in any control analyst's tool-kit. But by limiting the closed loop order of dynamics to the same order as the basic system, it trades simplicity for flexibility. The conventional control analyst has found the advantage of having additional dynamic elements available for shaping the frequency response characteristics over the whole frequency range so valuable in the past that he is reluctant to lose this design flexibility.

Other approaches to the multivariable design problem which retain this flexibility, although they are formulated in the frequency domain rather than in thé state space terms are the inverse Nyquist array technique of Rosenbrock [15], the characteristic Loci approach of MacFarlane [15] and the more recent procedures being developed by Horowitz and his co-workers [10]. It is surprising that these frequency domain approaches have received so little attention in the economic literature since they clearly have some relevance to economic control problems. For example, most econometric models are formulated and estimated in the observation-space rather than the state-space and the frequency-domain vectors are ideally suited to such applications. Control system design could then be based on a multivariable transfer function model in the observation-space, this MTF model is directly related to the final equations of Tinbergen (See Wallis and Prothero [28]).

The discussion above relates to the process of control design at some period prior to its application in the real world. At the design stage, exact information on the future real outcome is obviously not available but we can infer the likely properties of the real world by reference to the ensemble properties of the estimated model. Thus, in designing for robustness, we initially treat the nominal model as if it were an approximation to the real world, and then assess the sensitivity of the control design based on this nominal model by reference to its ensemble characteristics generated by varying aspects of its structure. In this sense, the ensemble of models thus obtained is utilised to represent the potential real world situations for the purpose of analysis. Another alternative would be to attempt an 'on-line' control of the economy and to adjust the control design in response to the deviation of the actual outcome from the desired path. Note that this procedure is not necessarily the same as currently suggested adaptive control schemes; it would however be equally difficult to implement in the economic situation, given the delay in true observations becoming available. It would in any case require the above analysis as a prelude to implementation.

4. A SIMPLE ILLUSTRATIVE EXAMPLE

In order to illustrate some of the points raised in previous sections of the paper, we consider a simple example discussed previously by Chow [4] in which annual government expenditure (u_t) is considered as the sole policy instrument to control annual gross national product (GNP) (y_t). Chow's analysis is based upon data for the US economy over the 20-year period 1953–72.

He first uses simple least squares regression analysis to estimate an equation in which y_t is related linearly to y_{t-1} and u_t and obtains the following result:

$$\underset{(0.072)}{y_t = 0.890\, y_{t-1}} + \underset{(0.311)}{0.779\, u_t} \qquad R^2 = 0.998,\quad S^2 = 159.1,\quad DW = 2.109 \qquad \text{Model A}$$

where the numbers in parentheses are asymptotic standard errors. In dynamic terms Model A has a steady state gain (long-run multiplier) of $0.779/(1-0.89) = 4.5$ and a time constant of $-(1/ln\ 0.89) = 8.6$ years respectively. (See R. G. D. Allen, *Mathematical Economics*.) This dynamic model is used as a basis for a comparative study in which the optimal LQG "certainty equivalent" solution is compared with solutions in which the uncertainty of the system is taken into account to differing degrees in the computation of alternative optimal control strategies.

Following Chow we exploit the obvious simplicity of the system to illustrate various fundamental aspects of control analysis which could well be obscured if we considered a more complex problem. However, the lessons learnt from this process will have implications for more realistic model structures.

Model Identification and Estimation

Bearing in mind our comments in section 2, it is interesting to note the estimation results obtained under different conditions and consider the implication of results on possible control system design.

The 'regression' model of Chow is based on the 'raw' data, the parameter estimates may be biased for various reasons, including those stressed in section 2.2 because of the problem of errors in variables, and also distorted because the simple linear model is not being applied to a small perturbation context. This is borne out by the results obtained when a recursive instrumental variable method (see [32], [29]) is applied to the data both in its raw form and after differencing both input series to remove the greater part of the pronounced trend evident in the raw data.

In the first case, the estimated model takes the form

$$\underset{}{y_t} = \underset{(0.05)}{0.02\, y_{t-1}} + \underset{(0.2)}{4.5\, u_t} \qquad \text{Model B}$$

which on using differenced data becomes

$$y_t = \underset{(0.04)}{0.57\, y_{t-1}} + \underset{(0.44)}{1.92\, u_t} \qquad \text{Model C}$$

Although these models look superficially different, we see on closer examination that they are quite similar in transfer function form: Model B is, effectively, a *static* model with $y_t \simeq 4.5\, u_t$ while Model C is a dynamic model with the same gain $(1.92/(1-0.57) = 4.5)$ as Model B but with a time constant of $-(1/\ln 0.57) = 1.8$ years. Bearing in mind the additional information on short-term dynamic behaviour imparted by the differencing operation, these results make reasonable sense in estimation terms, and the multiplier of 4.5 together with the time constant of 1.8 years in Model C also seem more reasonable in economic terms than those of the Chow model. On the other hand, the fit between the Model C output and the differenced data, is unremarkable and provides little confidence in the characterisation. In addition, recursive estimation of the two parameters as shown in Figure 5.1 gives evidence of parametric non-stationarity with the time constant increasing from 0.9 to 1.8 years over the 20-year period and the steady state gain oscillating as shown from 4.7 to 2.0 to 4.8 to 3.1 and finishing up at 4.4. The statistical significance of these parameter movements is open to question given the small sample size but the pattern of variation is clearly important.

From the above results, it would appear that the model is poorly identified in some manner: in other words, the model structure and possibly the input-output signal topology are inappropriate to the data as supplied. On the basis of the above discussion, the level of misspecification is likely to be larger than that implied by the standard error on the parameters (and the other 'specification' statistics) as

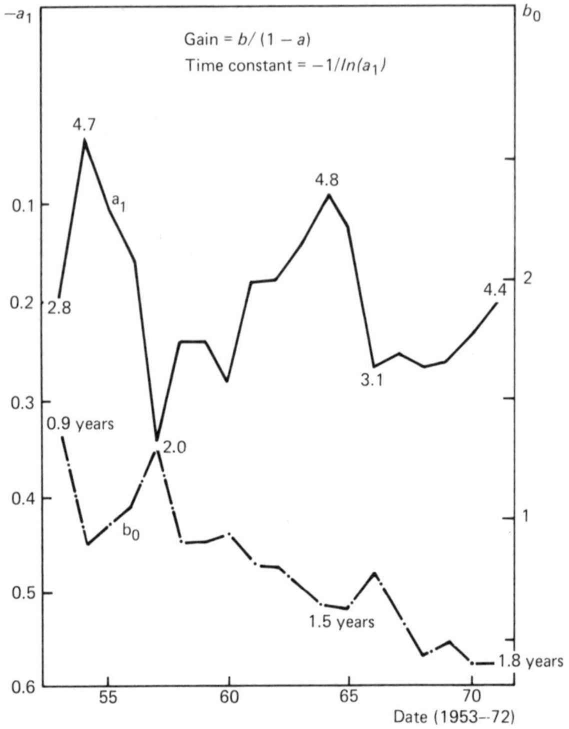

FIG. 5.1 Recursive Estimation Results

estimated by Chow. Thus if the model is to be used, as in this case, for control analysis, it will be necessary to take care that the resulting control system is 'robust' in the face of model misspecification.

Control Analysis

Chow uses Model A as a basis for optimal control calculations in which the aim is to steer GNP and government expenditures towards specified target paths (6 per cent annual growth of GNP and 5 per cent annual growth of government expenditures from their 1972 figures) whilst minimising a quadratic welfare cost function. The resulting control law is of the usual state variable feedback form

$$x_t = G_t y_{t-1} + g_t \tag{20}$$

where using Chow's nomenclature G_t is the state variable feedback gain and g_t is an 'intercept' term which, in effect, acts as an input command signal and helps to guide the system along the specified trajectories. Model A, together with equation (20), constitute a description of the closed loop system which Chow discusses mainly from the standpoint of the relative magnitude of the control parameters G_t and g_t obtained for the three optimal control formulations which he considers.

In order to consider factors such as the sensitivity of the controlled response to uncertainty and the implications of this on the 'robustness' of the optimal control laws, it is convenient and instructive to consider target paths rather simpler than those used by Chow. In particular, we will suppose that it is desired to *increase* GNP by an arbitrarily small amount (say 1 million dollars although this is clearly unimportant since we assume linearity) from the long-term trajectory, with the accompanying target path for the *change* in government spending being zero. In this way, we can assess the small perturbation 'step response' of the system and so evaluate the response characteristics rather more easily than if we were confronted by input and output both rising exponentially, as in Chow's example. In effect, his specification of exponentially growing target levels tends to obscure the short-term dynamic response characteristics which are so important to control system evaluation.

The control law is, of course, once again in the form given by (20) with the control gains as specified in Table 5.1 for a 10-year period as in Chow's example. The simulated response of the closed loop system with this control law operative is given in Figure 5.2(a): here the system is simulated by the model given in Model A with the nominal value of the parameters (i.e. 0.89 and 0.779). We refer to this result, therefore, as the

TABLE 5.1 Optimal Control Option
($A = 0.890$, $C = 0.779$, step response $T > 0$, $Y_d = 1$).

Period	*Gain*	*Intercept*
1	−0.57227	0.713
2	−0.57227	0.713
3	−0.57227	0.713
4	−0.57226	0.713
5	−0.57223	0.713
6	−0.57205	0.713
7	−0.57112	0.712
8	−0.56645	0.708
9	−0.54307	0.684
10	−0.43147	0.573

Note: We are considering the certainty equivalence solution.

'nominal' optimal response of the closed loop system. The response obtained when G_t and g_t are set at their 'steady state' values of $G_t = 0.572$ and $g_t = 0.713$ respectively is almost exactly the same as that shown in figure 5.2(a), indicating that time-variable gains are rather an unrequired luxury in this case.

A block diagram of the system in the simple optimal case is shown in Figure 5.3(a) where z is once again the forward shift operator (or z transform operator). Here y_{dt} can be considered as the target level; in this case, unity. Also shown in the figure are input and output disturbances. The exact nature of these disturbances is difficult to specify: for example, ξ_t could arise, as we have discussed in the previous section, from those other variables that effect GNP but do not appear in the model; similarly ζ_t could arise from similar considerations or be representative of observational error on GNP. The point is that the control analyst should, if possible, design the system to be robust in the face of all the inputs and parameter changes he could reasonably expect to occur. In this case, to define these simply in terms of the results of the rather naive model would clearly be insufficient and we must utilise his knowledge of the wider economic system to inject other factors not specifically uncovered by the rather limited estimation analysis.

Figure 5.2(a) also shows the responses obtained when uncertainty on the input parameter (the coefficient of u_t) is taken into account: the two-step responses shown here are obtained with exactly the same control law as that used for the nominal optimal response but with the input parameter in the simulated system model set to upper and lower bounds defined by the nominal value (0.779) plus and minus 0.31, which happens to be the standard error. We see that, in the face of this model misspecification, the control law is unable to generate a government spending policy which ensures that the target level is achieved and an undetected steady state offset from the target results. As we have pointed out in section 3 the reason for this lies in the form of the control law: in effect, control action is achieved in two stages; the feedback gain G_t principally governs the transient response of the system, controlling as it does the closed loop eigenvalue location; the intercept term is then chosen so that, with this value of G_t and the *nominal* values of the system parameters, the target level of GNP is achieved following the transient response. But the target level is not achieved if there is error on the estimated values of the parameters. We can see similar steady state offsets introduced if non-zero disturbances are introduced into the system: Figure 5.2(a), for example, also shows the responses obtained when the model parameters are set at their nominal values but with ξ_t set to a

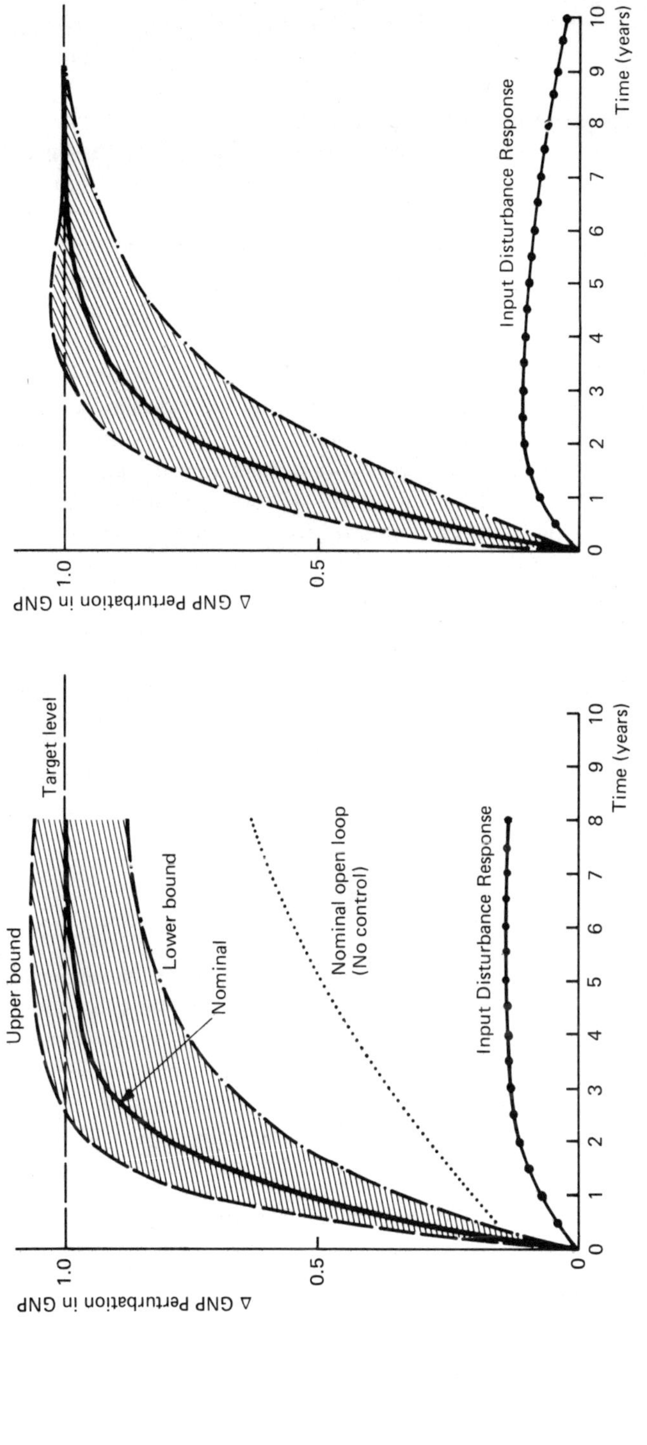

FIG. 5.2 Comparison of Optimal and Conventional Closed-loop Responses in the Presence of Model Misspecification

(a) Optimal SVF
Gain Margin = 4.3

(b) Conventional (Prop + Int + Pa)
Gain Margin = 3.6

constant 0.1 value (for clarity, g_t is set to 0 in the simulation, implying a target perturbation value of 0, i.e. the system is instructed to stay where it is and the GNP only changes because of the disturbances).

A block diagram of the modified control system with integral action is shown in Figure 5.3(b), which can be compared with that shown in Figure 5.3(a). Following the notations in section 3.2 we see now that the system is characterised by two state variables, y_t and z_t, where z_t is the integral of error between y_t and y_{dt}. In consequence, there are two state variable feedback gains, k_1 and k_2 (as yet unspecified) operating on these two state variables. It will be noted that the SVF system shown in this figure is simply a 'proportional plus integral' control system of the type discussed previously by Phillips. We also see that, in this constant target level context, it can be compared directly with the optimal control structure of Chow but, with suitable choice of k_1 and k_2 to ensure satisfactory stable transient response, it is *always* able to achieve the specified target level in the steady state despite any error (that is misspecification) in the model parameters. Moreover, as we have pointed out in previous sections, it is *always* possible to arbitrarily assign the eigenvalues of the closed loop system (and thus specify the closed loop dynamic modes of behaviour) by suitable choice of k_1 and k_2, provided the system remains linear and is controllable.

This type of control system design is demonstrated quite easily here since the closed loop transfer function (obtained by simple block diagram manipulation) is simply

$$\frac{y_t}{y_{dt}} = \frac{0.78k_2 z}{(z-1)(z-0.89+0.78k_1)+0.78k_2 z}$$

so that the closed loop characteristic equation is

$$z^2+(0.78k_1-1.89+0.78k_2)z+(0.89-0.78k_1)=0 \tag{26}$$

It is clear that if we specify a 'desired' characteristic equation whose roots are the required eigenvalues, say $d(z) = z^2 + d_1 z + d_2 = 0$, then the gains k_1 and k_2 which will nominally achieve the desired closed loop dynamics can be obtained by the solution of the linear simultaneous equations which result from equating similar powers of z in (26) and $d(z)$. For example, the eigenvalue of Chow's optimal closed loop system lies at 0.44, thus if we choose $d(z)$ to be $z(z-0,\ 44) = z^2 - 0.44z$, yielding $k_1 = \dfrac{0.89}{0.78}$ and $k_2 = \dfrac{0.56}{0.78}$, then the nominal transient response of the controlled system, shown in Figure 5.2(a), will be the same as that of the nominal transient response in the optimal design. But, of course, we will now have achieved the additional aim of zero steady state error between the GNP and its target value.

The response of the system is shown in Figure 5.2(c) for the nominal, upper and lower bound values on the simulated system parameters which can be compared directly with the optimal responses of Figure 5.2(a). We see that the transient response of the controlled system shown in Figure 5.2(c) is affected deleteriously by the error between the model and system parameters, as in the 'optimal' case, but regardless of such error the desired steady state is always achieved (unless such error induces closed loop instability).

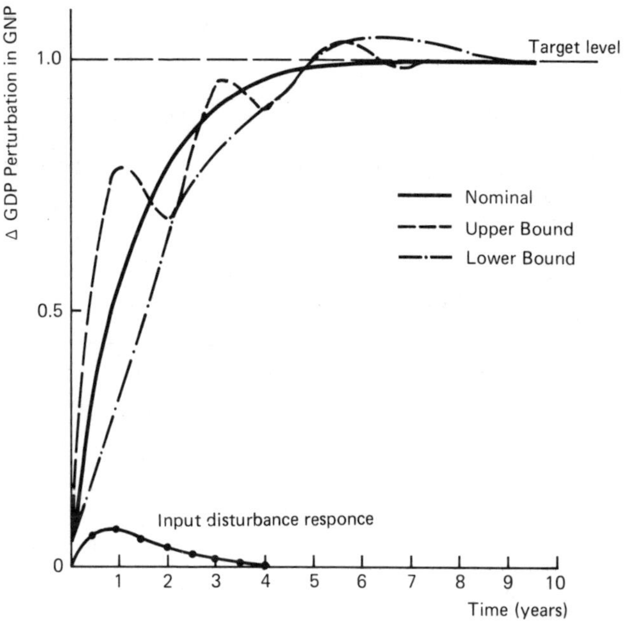

FIG. 5.2 Conventional PI System with Model misspecification
(c) Conventional (Prop + Int)
Gain Margin = 2.1

Of course, the simple PI controller of Figure 5.3(b) is still far from satisfactory in total control system terms. We have not, for example, considered at all either *how much* parametric error can be tolerated before it becomes unstable, or the nature of its response to the various kinds of disturbance, both deterministic and stochastic, that might be met in the real world.

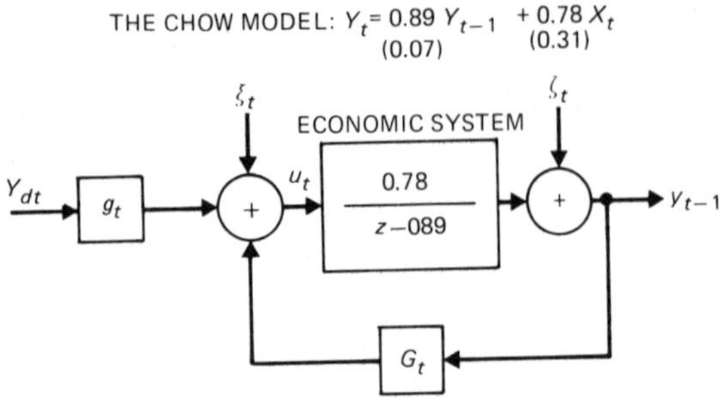

(a) Optimal SVF

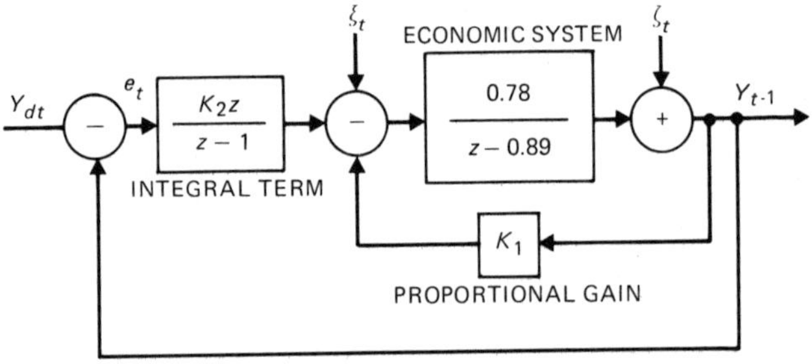

(b) SVF With Integral Action

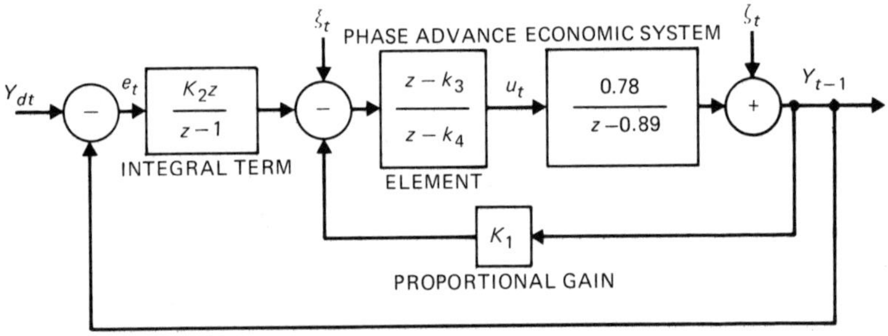

(c) SVF With Integral Action and Phase Advance
(i.e. conventional control)

FIG. 5.3 The Emerging Control System: from Optimal to Conventional?

In order to explore these questions more fully, it would be necessary to examine the full ensemble properties which is not possible here due to limited space (see [22]). It will suffice to say that improved 'phase and gain margins' (i.e. the margins in feedback loop gain and phase lag which exist before the onset of instability) and superior disturbance response can be achieved, as we have pointed out in general terms earlier, by the addition of dynamic elements in the control law. For instance, in the present case a simple example would be the addition of a phase advance filter as shown in Figure 5.3(c), which yields the closed loop response characteristics shown in Figure 5.3(c); these are certainly an improvement over the simple PI system, with less tendency to instability at the lower bound but still acceptable transient and steady-state behaviour.

Although the control design shown in Figure 5.3(c) is still not fully satisfactory, this example illustrates the manner in which the frequency response characteristics of the closed loop system may be carefully tailored to produce the desired overall closed loop performance which will be a compromise between many conflicting demands of the economic policy analyst. (In this single input-single output example, the interested reader unacquainted with this conventional approach to control system design is referred to Truxal [24], who discusses all aspects of conventional design in considerable detail.)

Such a compromise is necessary for the *practical* solution to any control problem and cannot be achieved by the blind application of *any* control theoretic methodology. Thus we must look for new procedures which more nearly achieve our overall aims. In the present context, for example, one possible improvement would be to combine the optimal and conventional approaches in the manner suggested in [30] and [2], [13]. But these are certainly not the only nor necessarily the best ways of proceeding; the problem is difficult and a simple solution is unlikely.

Finally, it should be emphasised again that we view the kind of control system design procedures discussed above as being appropriate only for 'small-perturbation' control *unless* a linear description is considered appropriate for large-perturbation (in the present context, long-run) behaviour. In other words, we have effectively decomposed the problem into short and long period 'sub-systems' (high and low frequency behaviour) and, in effect, considered only the control of the short period perturbations about some specified longer-term target trajectory. In this manner, we can be reasonably confident that the linear analysis is applicable.

We have not discussed here the specification or nature of the long-term target behaviour, but we assume that it would be generated either by policy discussion relating to the long term, optimisation studies or a combination of both [18]. In all cases, however, a 'long-term model' of some sort would be required as a basis for study. In the present rather artificial and simplistic problem, an appropriate long-term characterisation would appear to be a stochastic form of Model B; namely the simple steady-state relationship

$$y_t = \underset{(0.2)}{4.5}\, u_t + e_t$$

where e_t represents stochastic disturbances to the system and the standard error on the coefficient is retained to emphasise the uncertainty in the parameterisation. The trajectories resulting from the long-term analysis, say y_t^* and u_t^* would then provide the basis for a definition of the small perturbation variables, i.e.

$$\delta y_t = y_t - y_t^*;\ \delta u_t = u_t - u_t^*$$

NOTES

1. Clearly the size of most practical forecasting models inhibits their utility in control terms; more problems being created by the excessive and often spurious detail than are solved. A recent study of the structure of ten such models of various sizes, [21], has shown that much of the structure of these models results simply from a desire to generate more output information rather than from any need to increase the internal explanatory potential of the model. It is knowledge of these internal functions of the economic system which is of vital importance for control system design.
2. Since this paper was written reference [5] has appeared in the literature: it approaches these questions but from a different standpoint.

REFERENCES

[1] Anderson, B. D. O., and Moore, J. B., *Linear Optimal Control* (Prentice-Hall, 1971).
[2] Athans, M., 'On the Design of P. I. D. Controllers Using Optimal Linear Regulator Theory', *Automatica*, vol. 7 (1971).
[3] Beck, M. B. and Young, P. C., 'Systematic Identification of DO-BOD Model Structure *Journal of the Environmental Engineering Division, A S C E*, vol. 102, no. EE5, pp. 909–27 (Oct 1976).
[4] Chow, G. S., *Analysis and Control of Dynamic Economic Systems* (Wiley, New York, 1975).
[5] Chow, G. S., 'Usefulness of Imperfect Models for the Formulation

of Stabilization Policies', *Annals of Economic and Social Measurement*, vol. 6, no. 2 (1977).

[6] Diestler, M. and Seifert, H. G., 'Identifiability and Consistent Estimability in Econometric Models', mimeo (1975).

[7] Davison, E. J., 'Multivariable Tuning Regulators: The Feedforward and Robust Control of a General Servomechanism Problem, *IEEE Trans. Aut. Control*, vol. AC21, no. 1, pp. 35–47 (Feb 1976).

[8] Fuller, A. T., 'Feedback Control Systems with Low Frequency Stochastic Disturbances', University of Cambridge, Dept. of Engineering, CUED/F Control/TR 97 (1975).

[9] Gilbert, E. G., 'The Decoupling of Multivariable systems by Means of State Variable Feedback', *SIAM Jnl. Control*, vol. 7, pp. 50–63 (1969).

[10] Horowitz, I. M., and Shaked, U., Superiority of Transfer Function over State-Variable Methods in Linear Time-Invariant Feedback System Design, *IEEE Trans. Aut. Control*, vol. AC20, no. 1, pp. 84–97 (Feb 1975).

[11] Johnson, C. D., 'Further Study of the Linear Regulator with Disturbances', *IEEE Trans. Aut. Control*, vol. AC15, no. 2, pp. 222–8 (Apr 1970).

[12] Johnson, C. D., 'Accommodation of External Disturbances in Linear Regulator and Servomechanism Problems', *IEEE Trans. Aut. Control*, vol. AC16, no. 6, pp. 635–44, (Dec 1971).

[13] Johnson, C. D., '*Accommodation of Disturbances in Optimal Control Problems*', *Int. Jnl. of Control*, vol. 15, no. 2 (1972).

[14] Mehra, R. K., 'Identification in Control and Econometrics: Similarities and Differences', *Annals of Economic and Social Measurement*, 3/1 (1974).

[15] MacFarlane, A. G. J., 'A Survey of Some Recent Results in Linear Multivariable Feedback Theory', *Automatica*, vol. 8, pp. 455–92 (1972).

[16] Phillips, A. W., 'Stabilization Policy in a Closed Economy', *Economic Journal*, June 1954.

[17] Phillips, A. W., 'Stabilization Policy and the Time Forms of Lagged Responses', *Economic Journal*, June 1957.

[18] Pierce, D., 'Quantitative Analysis for Decisions at the Federal Reserve, *Annals of Economic and Social Measurement*, vol. 3, no. 1 (1974).

[19] Preston, A. J., and Wall, K. D., 'Some Aspects of the Use of State Space Models in Econometrics', IFAC/IFORS Conference, Warwick, 1973.

[20] Porter, B. and Crossley, R., *Modal Control* (Taylor & Francis, London, 1972).

[21] Salmon, M. H., 'Structural Properties of Econometric Models', CRES Working Paper RG/16 (ANU, 1977).

[22] Salmon, M. H., and Young, P. C., 'Some Aspects of Control Methods and Robust Economic Policy', CRES Report no. AS/R12 (1977).

[23] Salmon, M. H., 'Recursive Formulations for Econometric Estimators', CRES Working Paper RG/17 (ANU, 1977).

[24] Truxal, J. G., *Automatic Feedback Control System Synthesis*, (McGraw-Hill, New York, 1955).

[25] Unbehauer, H., Schmid, C., and Bottiger, F., 'Comparison and Application of DDC Algorithms for a Heat Exchanger', *Automatica*, vol. 12, no. 5, pp. 293–402 (Sep 1976).

[26] Wang, S. H., and Desoer, C. A., 'The Exact Model Matching of Linear Multivariable Systems', *IEEE Trans. Aut. Control*, vol. AC17, pp. 347–8 (June 1972).

[27] Wonham, W. M., *Linear Multivariable Control* (Springer Verlag, Lecture Notes in Economics and Mathematical Systems no. 101, 1974).

[28] Wallis, K. F., and Prothero, D. L., Modelling Macroeconomic Time Series, *Jnl. R. Statistical Soc.*, series A, vol. 139 (1976).

[29] Young, P. C., '*Recursive Approaches to Time Series Analysis*', *Bulletin of the Institute of Mathematics and its Applications*, vol. 10, nos 5 and 6 (May/June 1974).

[30] Young, P. C., and Willems, J. C., 'An Approach to the Linear Multivariable Servomechanism Problem', *International Journal of Control*, vol. 15, no. 5 (1972).

[31] Young, P. C., Sheswell, S. H., and Neethling, C., 'A Recursive Approach to Time-Series Analyses', Control Div. Department of Engineering, University of Cambridge, Tech. Report no. CUED/B-Control/TR16 (1971).

[32] Young, P. C., 'Some Observations on Instrumental Variable Methods of Time-Series Analysis', *Int. Jnl. of Control*, vol. 23, no. 5, pp. 593–612 (1976).

[33] Young, P. C., and Levsen, L. D., *Linear System Design: Application of State Space Methods to the Design of Linear Time Invariant Systems* (A Handbook to a Package of FORTRAN V Computer Programs), Naval Weapons Center, China Lake, California, Publication No. NWC-TP-4927.

[34] Young, P. C., and Yancey, C., *A Second Generation Adaptive*

Autostabilisation System for Aircraft and Missiles, Naval Weapons Center, China Lake, California, Report No. TN-404-109 (1975).

[35] Young, P. C., 'A General Theory of Modelling for Badly Defined Systems', to appear in *Modelling of Land, Air and Water Resource Systems*, ed. by G. C. Vansteekiste (Academic Press, 1977).

[36] Young, P. C., and Jakeman, A. J., *Refined Instrumental Variable Methods of Recursive Time-Series Analysis, Part I: Single Input-Single Output Systems*, CRES Report No. AS/R12 (1977).

[37] Jakeman, A. J., and Young, P. C., *Refined Instrumental Variable Methods of Recursive Time-Series Analysis, Part II: Multivariable Systems*, CRES Report No. AS/R13 (1977).

6 Iterative Respecification of the Quadratic Objective Function

B. Rüstem, J. H. Westcott, M. B. Zarrop,
S. Holly and R. Becker

1. INTRODUCTION

A fundamental problem in the optimisation of policy decisions is the specification of a suitable objective function. It has been argued by Rustem, Velupillai and Westcott (1976) and Westcott, Holly, Rustem and Zarrop (1976) that quadratic functions penalising the weighted deviation of the computed trajectories from their desired values form an acceptable class of objective functions. An iterative method for specifying the weighting matrices of such quadratic functions has been reported by Rustem, Velupillai and Westcott (1976). The method is not concerned with a 'best' set of weights *independent* of the 'best' desired path; rather a politically acceptable path, optimally generated, is the main aim. This paper discusses the numerical considerations arising from the actual implementation of the algorithm described by Rustem, Velupillai and Westcott (1976). Basic concepts and the method are introduced in Sections 2 and 3 using formal mathematical definitions parallel to an intuitive approach to the underlying problem. Section 4 describes the computational aspects of the implementation of the algorithm and Section 5 illustrates the method giving numerical examples.

2. PRELIMINARIES

A quadratic objective function can be considered equivalent to the second order Taylor series expansion about the unconstrained optimum of the true nonlinear objective function. The constant term of this expansion can be ignored without affecting the value of the optimal trajectory. Consider, therefore, the quadratic function

$$J(\underline{Y}, \underline{U}) = \frac{1}{2} < \left(\begin{bmatrix} \underline{Y} \\ \underline{U} \end{bmatrix} - \begin{bmatrix} \underline{Y} \\ \underline{U} \end{bmatrix}^d \right), \tilde{Q} \left(\begin{bmatrix} \underline{Y} \\ \underline{U} \end{bmatrix} - \begin{bmatrix} \underline{Y} \\ \underline{U} \end{bmatrix}^d \right) > \quad (2.1)$$

where $\tilde{Q}$ is a symmetric $(n+m)N \times (n+m)N$ matrix,

$$\underline{Y}^T \triangleq [\underline{y}^T(1), \ldots, \underline{y}^T(k), \ldots, \underline{y}^T(N)] \quad (2.2)$$

is the mN-vector of the trajectory of the endogenous variables in the time period $[1, N]$ with the m-vector $\underline{y}(k)$, $1 \leqslant k \leqslant N$, denoting the endogenous variables at time k,

$$\underline{U}^T \triangleq [\underline{u}^T(1), \ldots, \underline{u}^T(k), \ldots, \underline{u}^T(N)] \quad (2.3)$$

is the mN-vector of the trajectory of policy instruments in the time period $[1, N]$ with the m-vector $\underline{u}(k)$, $1 \leqslant k \leqslant N$, denoting the policy instruments at time k. The superscript d on $\underline{Y}$ and $\underline{U}$ denotes the desired trajectories or targets. The desired trajectories are also the minima of the Taylor series expansion (2.1). Since this expansion is about the unconstrained minimum of the true nonlinear objective function, $\underline{Y}^d$ and $\underline{U}^d$ are exactly the values of $\underline{Y}, \underline{U}$ which minimise the true welfare function. The vectors $\underline{Y}^d$ and $\underline{U}^d$ may be thought of as desired but possibly infeasible values considering the restriction imposed by the constraints of the system. The constrained optimisation problem that will next be considered can be interpreted as getting as near as possible to these desired values (see Remark (2.8)).

Definition (2.4)

The optimisation problem involved in the minimisation of (2.1) subject to the restrictions of the model equations as well as any other constraints will be defined as

$$\min\{J(\underline{Y}, \underline{U}) \mid \underline{Y}, \underline{U} \in \Omega, \underline{h}(\underline{Y}, \underline{U}) = \underline{0}, \underline{g}(\underline{Y}, \underline{U}) \geqslant \underline{0}\} \quad (2.5)$$

The vector valued function $\underline{h}$ denotes the model equations as well as any other equality constraints imposed on $\underline{Y}$ and $\underline{U}$. The vector function $\underline{g}$ denotes the inequality constraints on the values of $\underline{Y}$ and $\underline{U}$. The set constraint $\Omega \subset E^{(n+m)N}$ is the set of admissible values of $\underline{Y}, \underline{U}$ from the policy maker's point of view.

Definition (2.6)

The set of $\underline{Y}, \underline{U}$ satisfying the functional constraints of (2.5) is defined as

$$F \triangleq \{\underline{Y}, \underline{U} \in E^{(n+m)N} \mid \underline{h}(\underline{Y}, \underline{U}) = \underline{0}, \underline{g}(\underline{Y}, \underline{U}) \geqslant \underline{0}\} \quad (2.7)$$

Remark (2.8)

Under the assumption of convexity of F (e.g. when $\underline{h}$ and $\underline{g}$ are linear) the solution to

$$\min\{J(\underline{Y}, \underline{U}) \mid \underline{Y}, \underline{U} \in F\} \quad (2.9)$$

may be interpreted as the point on F nearest to $\underline{Y}^d, \underline{U}^d$ with respect to the norm

$$\left\| \begin{bmatrix} \underline{Y} \\ \hline \underline{U} \end{bmatrix} - \begin{bmatrix} \underline{Y} \\ \hline \underline{U} \end{bmatrix}^d \right\|^2_{\tilde{Q}} \quad (2.10)$$

for positive definite $\tilde{Q}$. Hence the solution to (2.9) is the projection of the desired trajectories $\underline{Y}^d$, $\underline{U}^d$ on to F.

Assumption (2.11)

The feasible set of the optimisation problem (2.5) is not empty, i.e.

$$F \cap \Omega \neq \phi.$$

This simple assumption is central to the discussion in subsequent sections. Points in this intersection have to be sought by solving (2.9) for different $J(\underline{Y}, \underline{U})$. Thus (2.9) has to be solved a number of times by respecifying $J(\underline{Y}, \underline{U})$ until a solution $[\underline{Y}^T \vdots \underline{U}^T]^*$ is found such that

$$\begin{bmatrix} \underline{Y} \\ \hline \underline{U} \end{bmatrix}^* \in F \cap \Omega \quad (2.12)$$

It should be noted that in contrast to the algebraic equations describing F, the subspace Ω exists only in the mind of the rational policy maker.

Remark (2.13)

Dynamic optimisation problems are included in formulation (2.5). The correspondence between dynamic and static optimisation problems is extensively discussed in Canon, Cullum and Polak (1970) and Polak (1971). Definitions (2.2) and (2.3) for $\underline{Y}$ and $\underline{U}$ have been made to suit such multi-time-period decision problems.

Given the desired values $[\underline{Y}^T \vdots \underline{U}^T]^d$, and the region F which will be assumed to be convex from now on, the nearness of the solution of (2.9) to $[\underline{Y}^T \vdots \underline{U}^T]^d$ is affected only by the weighting matrix $\tilde{Q}$ in (2.1). Different values of this matrix will therefore define different points on F as the nearest point to $[\underline{Y}^T : \underline{U}^T]^d$. Thus, given these desired trajectories the only way of respecifying $J(\underline{Y}, \underline{U})$ to produce an optimal solution in $F \cap \Omega$ is to respecify $\tilde{Q}$.

Let $\tilde{Q}_c$ denote the current weighting matrix of (2.1). The corresponding solution of (2.9) using $\tilde{Q}_c$ will be called 'the current optimal solution',

$$[\underline{Y}^T : \underline{U}^T]_c^* \tag{2.14}$$

After a careful inspection of the values of (2.14), the policy-maker may decide that some of these values are not quite what he wants them to be. This means (2.14) does not satisfy his set constraint, i.e.

$$[\underline{Y}^T : \underline{U}^T]_c^* \notin \Omega \tag{2.15}$$

One way[1] of obtaining an optimal solution which the policy-maker may prefer to the current one (2.14) is to alter some of the elements of the weighting matrix $\tilde{Q}_c$. The basic idea is to update $\tilde{Q}_c$ to a new weighting matrix denoted by $\tilde{Q}_n$ such that 'the new optimal trajectory'

$$[\underline{Y}^T : \underline{U}^T]_n^* \tag{2.16}$$

obtained by solving (2.9) with $\tilde{Q}_n$ is also in Ω (see Figure 6.1 below). The simple conjecture that the new optimal solution (2.16) may not be completely satisfactory (i.e. not in Ω) as a result of the first update of $\tilde{Q}$ leads to an iterative procedure updating $\tilde{Q}$ and solving the corresponding optimisation problem until the solution which is also in Ω is obtained.

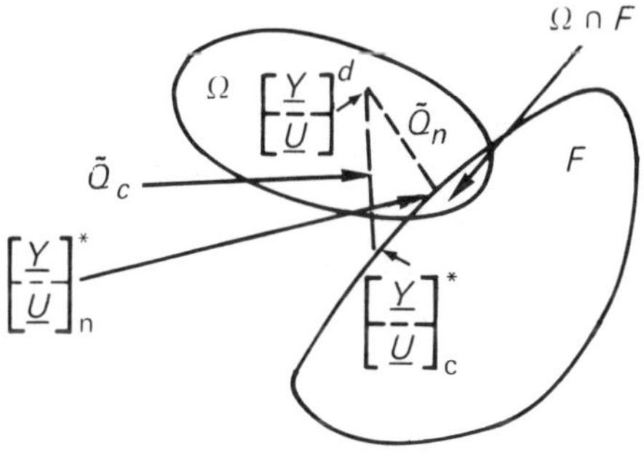

FIG. 6.1

For the purpose of computing the updates of $\tilde{Q}$, the policy-maker is required to specify the trajectory he prefers to the current optimal solution (2.14). Hence the 'preferred' trajectory

$$[\underline{Y}^T : \underline{U}^T]^P \tag{2.17}$$

incorporates all the corrections to (2.14) such that (2.17) is preferable to the 'current' optimal trajectory. An obvious property of the preferred trajectory is that it is in Ω. The weighting matrix $\tilde{Q}_c$ has thus to be updated such that the next optimal solution (2.16) is in the neighbourhood of (2.17) (see Lemma (4.3)). Given the 'current' optimal trajectory (2.14) and the 'preferred' trajectory (2.17) the displacement (or correction) vector is defined as

$$\underline{\delta} \triangleq \begin{bmatrix} \underline{Y} \\ \underline{U} \end{bmatrix}^P - \begin{bmatrix} \underline{Y} \\ \underline{U} \end{bmatrix}^*_c \tag{2.18}$$

Let $J_{\text{true}}(\underline{Y}, \underline{U})$ and $\nabla J_{\text{true}}(\underline{Y}, \underline{U})$ denote respectively the true nonlinear objective function and its gradient. This true objective function need not be seen as a purely abstract concept. If it is taken to be a quadratic, its weighting is what is actually being sought (see e.g. Theorem (3.1)). The vector

$$\underline{\gamma} = \nabla J_{\text{true}}(\underline{Y}^P, \underline{U}^P) - \nabla J_{\text{true}}(\underline{Y}^*_c, \underline{U}^*_c) \tag{2.19}$$

is thus the difference of the gradients of J_{true} evaluated at (2.17) and (2.14). The weighting matrix which will give a new optimal solution (2.16) nearer to the preferred values may be computed by adding a rank one correction term to the current weighting matrix using the following formula

$$Q_n = Q_c + \frac{(\underline{\gamma} - Q_c\underline{\delta})(\underline{\gamma} - Q_c\underline{\delta})^T}{\langle \underline{\gamma} - Q_c\underline{\delta}, \underline{\delta} \rangle} \tag{2.20}$$

This updating formula is a member of the variable metric family widely used for computing approximations to second derivative matrices in unconstrained (see e.g. Broyden (1967), Davidon (1968), Murtagh and Sargent (1970) and constrained (Murtagh and Sargent (1969)) optimisation. The vector $\underline{\gamma}$ represents the change in the gradient vector of the true objective function evaluated at the preferred and at the current optimal trajectories. Clearly, assigning values to $\underline{\gamma}$ is not a trivial problem and is discussed in Section 4.

Remark (2.21)

In general, variables may be crossweighed intertemporally using the updating formula (2.20). For example, unemployment at time t may be weighed against the balance of payments at time $t + i$. This also applies to weighting $\underline{y}(t)$ against $\underline{u}(t + i)$.

Remark (2.22)

Objection of Economists to the Symmetric Nature of the Quadratic Objective Function

The quadratic objective function penalises symmetrically deviations of the optimal trajectories from the desired trajectories. Hence penalisation occurs irrespective of the possibility that the optimal trajectory may have a value even better than the desired trajectory. No distinction is made between such deviations and those when the optimal trajectory is inferior in value to the desired trajectory. This undesirable property of the quadratic objective function may be overcome in two ways. The first involves the use of piecewise quadratic functions and is discussed extensively by Friedman (1972). Alternatively, when the desired values are less acceptable than the current optimal values, one can view deviations from them as an indication of their *misspecification* and modify them accordingly. This is discussed further in Rustem, Velupillai and Westcott ((1976), Remark (6.12)).

3. THE BASIC CONCEPTUAL ALGORITHM

The updating formula (2.20) is used to adjust the weighting matrix Q of the quadratic objective function (2.1). The adjustments are made to generate optimal policies which the policy-maker may prefer to the current optimal trajectory (2.14). The process of adjusting the weights may be repeated until a satisfactory analysis is made and an acceptable optimal solution is attained. Figure 6.2 illustrates the iterative nature of the method. As in Figure 6.2, if the new optimal trajectory (2.16) is not

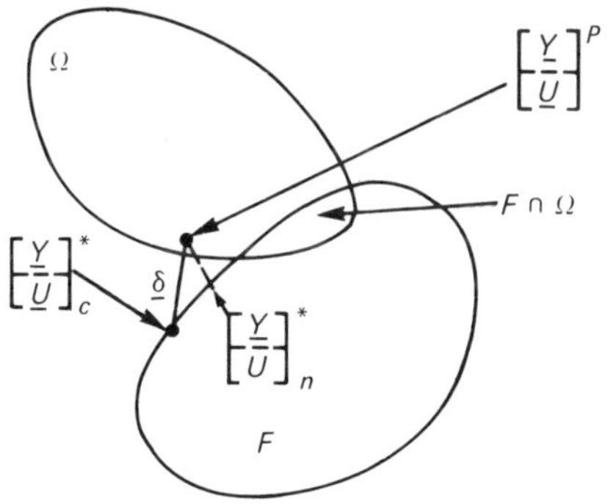

FIG. 6.2

quite in Ω, the procedure of adjusting Q has to be repeated. Conceptually, the resulting iterative procedure needs to follow the following steps:

Step 0:
Given the desired trajectory

$$\begin{bmatrix} \underline{Y} \\ \hline \underline{U} \end{bmatrix}^d$$

assume some initial weighting matrix $\tilde{Q}_c$.

Step 1:
Using the weighting matrix $\tilde{Q}_c$ and the desired trajectory in (2.1) solve the minimisation problem (2.9) to obtain the current optimal trajectory (2.14).

Step 2:
If, according to the policy maker

$$\begin{bmatrix} \underline{Y} \\ \hline \underline{U} \end{bmatrix}_c^* \in \Omega \text{ (and hence } \begin{bmatrix} \underline{Y} \\ \hline \underline{U} \end{bmatrix}_c^* \in \Omega \cap F) \text{ stop.}$$

Otherwise ask the policy-maker to specify the changes required in the current optimal trajectory to make it acceptable. The preferred trajectory (2.17) is thus specified along with the corresponding $\underline{\gamma}$ vector (2.19). Methods for specifying $\underline{\gamma}$ are discussed in Section 4.

Step 3:
Given the vector $\underline{\gamma}$, the current optimal and the preferred trajectories compute the displacement vector $\underline{\delta}$ (2.18) and using the updating formula (2.20) compute $\tilde{Q}_n$. Set $\tilde{Q}_c = \tilde{Q}_n$ and go to Step 1.

The denominator of the updating formula (2.20) must be safeguarded from becoming zero. This can be accomplished with small changes to $\underline{\delta}$ and/or $\underline{\gamma}$ within limits acceptable to the policy maker. Provided this is done, $\tilde{Q}_n$ remains positive definite if $\tilde{Q}_c$ is positive definite.

Theorem (Fiacco and McCormick (1968)): (3.1)

Motivation for Convergence
Assume the true objective function to be a positive definite quadratic form. Assume, also, that $(n+m)N$ linearly independent $\underline{\delta}$ vectors and corresponding values for $\underline{\gamma}$ may be specified during the iterative procedure. If no zero denominators occur in the updating formula then starting with any positive definite initial weighting matrix, the generated

updates converge to the weighting matrix of the true objective function after at most $(n+m)N$ iterations.

An important aspect of (3.1) is that the specified $\underline{\delta}$ and $\underline{\gamma}$ vectors need to be consistent with the true welfare function. Theorem (3.1) remains as only a motivation for the convergence of the iterative procedure precisely because it may be very difficult for the policy-maker to satisfy such requirements on $\underline{\delta}$ and $\underline{\gamma}$.

The effect of respecifying the weighting matrix on the optimal trajectories will now be investigated under the assumption of only linear equality constraints describing the region F.

Definition (3.2)

The convex region F (2.7) is assumed to be defined only by ℓ linearly independent equations and is redefined as the linear manifold

$$F(\underline{b}) \triangleq \{\underline{Y}, \underline{U} \in E^{(n+m)N} \mid H^T \begin{bmatrix} \underline{Y} \\ \overline{\underline{U}} \end{bmatrix} = \underline{b}\} \tag{3.3}$$

where H is an $\ell \times (m+n)N$ matrix and $\underline{b}$ an ℓ-vector. The ℓ columns on H are thus linearly independent and $\ell \leqslant (m+n)N$. The equations of the model are included in (3.3). The subspace of $F(\underline{b})$ is clearly $F(\underline{O})$.

As mentioned in (2.8) the current optimal trajectory obtained using $\tilde{Q}_c$ in (2.9) is the projection of the desired trajectories on to $F(\underline{b})$ with respect to the norm (2.10) and is given by

$$\begin{bmatrix} \underline{Y} \\ \overline{\underline{U}} \end{bmatrix}_c^* = \begin{bmatrix} \underline{Y} \\ \overline{\underline{U}} \end{bmatrix}^d - \tilde{Q}_c^{-1} H (H^T \tilde{Q}_c^{-1} H)^{-1} \left(H^T \begin{bmatrix} \underline{Y} \\ \overline{\underline{U}} \end{bmatrix}^d - \underline{b}\right) \tag{3.4}$$

Correspondingly, the projection operator

$$P = I - \tilde{Q}_c^{-1} H (H^T \tilde{Q}_c^{-1} H)^{-1} H^T \tag{3.5}$$

projects all vectors in $E^{(n+m)N}$ on to the subspace $F(\underline{O})$. The derivations of (3.4) and (3.5) are given by Rustem (1977).

Lemma (3.6)

If $\tilde{Q}_n$ is computed as the update of $\tilde{Q}_c$ using (2.20), the new optimal solution (2.16) may be expressed in terms of the current optimal solution and a correction term:

$$\begin{bmatrix} \underline{Y} \\ \overline{\underline{U}} \end{bmatrix}_n^* = \begin{bmatrix} \underline{Y} \\ \overline{\underline{U}} \end{bmatrix}_c^* + \alpha \underline{\beta} \tag{3.7}$$

where the scalar

$$\alpha = \frac{(\tilde{Q}_c^{-1}\underline{\gamma}-\underline{\delta})^T\tilde{Q}_c\left(\begin{bmatrix}\underline{Y}\\ \underline{U}\end{bmatrix}^*_c-\begin{bmatrix}\underline{Y}\\ \underline{U}\end{bmatrix}^d\right)}{(\tilde{Q}_c^{-1}\underline{\gamma}-\underline{\delta})^TH(H^T\tilde{Q}_c^{-1}H)^{-1}H^T(\tilde{Q}_c^{-1}\underline{\gamma}-\underline{\delta})-(\tilde{Q}_c^{-1}\underline{\gamma}-\underline{\delta})^T\underline{\gamma}} \tag{3.8}$$

and the vector

$$\underline{\beta} = P(\tilde{Q}_c^{-1}\underline{\gamma}-\underline{\delta}) \tag{3.9}$$

is the projection of $(\tilde{Q}_c^{-1}\underline{\gamma}-\underline{\delta})$ on to $F(\underline{O})$. Thus the correction to the current optimal solution lies along this projection.

Proof

The new optimal trajectory is obtained by replacing $\tilde{Q}_c$ with $\tilde{Q}_n$ in (3.4). The Sherman-Morrison formula may be used with (2.20) to express $\tilde{Q}_n^{-1}$ and $(H^T\tilde{Q}_n^{-1}N)^{-1}$ in (3.4) in terms of $\tilde{Q}_c^{-1}$ and $(N^T\tilde{Q}_c^{-1}N)^{-1}$. After some rearranging

$$\begin{bmatrix}\underline{Y}\\ \underline{U}\end{bmatrix}^*_n=\begin{bmatrix}\underline{Y}\\ \underline{U}\end{bmatrix}^*_c-\frac{(\tilde{Q}_c^{-1}\underline{\gamma}-\underline{\delta})^TH(H^T\tilde{Q}_c^{-1}H)^{-1}\left(H^T\begin{bmatrix}\underline{Y}\\ \underline{U}\end{bmatrix}^d-\underline{b}\right)}{\Delta(\underline{\gamma},\underline{\delta})}\underline{\beta} \tag{3.10}$$

is obtained where the scalar $\Delta(\underline{\gamma},\underline{\delta})$ denotes the denominator of (3.8). The relationship

$$\nabla J(\underline{Y}_c^*, \underline{U}_c^*) = H\underline{\lambda} \tag{3.11}$$

is given by Lagrange multiplier theory for an ℓ-vector $\underline{\lambda}$. Since the current optimal trajectory is feasible and hence lies on $F(\underline{b})$ given by (3.3) the equality

$$H^T\begin{bmatrix}\underline{Y}\\ \underline{U}\end{bmatrix}^*_c=\underline{b} \tag{3.12}$$

holds. Thus, using (3.12) and the fact that (3.11) denotes the gradient of the quadratic function (2.1)

$$\begin{aligned}H^T\begin{bmatrix}\underline{Y}\\ \underline{U}\end{bmatrix}^d-b &= -H^T\left(\begin{bmatrix}\underline{Y}\\ \underline{U}\end{bmatrix}^*_c-\begin{bmatrix}\underline{Y}\\ \underline{U}\end{bmatrix}^d\right)\\ &= -H^T\tilde{Q}_c^{-1}\nabla J(\underline{Y}_c^*, \underline{U}_c^*)\\ &= -H^T\tilde{Q}_c^{-1}H\underline{\lambda}\end{aligned} \tag{3.13}$$

is obtained. Substituting (3.13), the gradient of $J(\underline{Y}_c^*, \underline{U}_c^*)$ and (3.11) in (3.10) gives the required result.

4. COMPUTATIONAL CONSIDERATIONS FOR THE IMPLEMENTATION OF THE ALGORITHM

The main problem with the implementation of the algorithm discussed in Section 3 is the specification of $\underline{\gamma}$ (2.19). This vector measures the change in the gradient of the true objective function. In order to see what this change means, it is necessary to understand what the gradient of the objective function stands for. The gradient with respect to the endogenous variable $y_i(k)$ or policy instrument $u_j(k)$ at time k

$$\left.\frac{\delta J_{\text{true}}}{\delta y_i(k)}\right|_{y_i(k) = y_i^o(k)} \quad \text{or} \quad \left.\frac{\delta J_{\text{true}}}{\delta u_j(k)}\right|_{u_j(k) = u_j^0(k)}$$

evaluated at the fixed values $y_i^o(k)$, $u_j^0(k)$ denotes the 'priority' attached to a unit increase from $y_i^o(k)$ or $u_j^o(k)$. Thus $\underline{\gamma}$ denotes the change in these 'priorities' when going from the current optimal to the preferred trajectories. In Bray (1972) a way is described for quantifying these gradients by asking policy-makers direct questions about their priorities on different variables. This is similar, in spirit, to the propositions made by Frisch (1972). Numerical results obtained using such an approach are reported by Bray (1975). If $\underline{\gamma}$ is obtained in this way, Theorem (3.1) ensures the convergence, in finite number of steps, of the algorithm in Section 3 to the true weighting matrix of the true quadratic objective function. The requirement that at each iteration of the algorithm the specified $\underline{\delta}$ and $\underline{\gamma}$ vectors need to be consistent with the true objective function has to be fulfilled for this result to hold.

Apart from the difficulties of specifying a different $\underline{\gamma}$ at each iteration and of fulfilling the consistency requirement, a further complication arises when policy-makers prefer not to be asked questions about their priorities on different variables. An alternative and more mechanical way of choosing $\underline{\gamma}$ has been developed to overcome such problems. The basis of this method lies in the desirable form the new optimal solution is expressed in. The vector $\underline{\gamma}$ is taken as a scalar multiple of $\tilde{Q}_c\,\underline{\delta}$, i.e.

$$\underline{\gamma} = \phi \tilde{Q}_c \underline{\delta} \tag{4.1}$$

hence the updating formula (2.20) becomes

$$\tilde{Q}_n = \tilde{Q}_c + (\phi - 1)\frac{\tilde{Q}_c \underline{\delta}\underline{\delta}^T \tilde{Q}_c}{<\underline{\delta}, \tilde{Q}_c \underline{\delta}>} \tag{4.2}$$

Lemma (4.3)

If $\underline{\gamma}$ is given by (4.1) and hence $\tilde{Q}_n$ by (4.2) the correction term in (3.7) reduces to a scalar multiple of the projection of $\underline{\delta}$ on to $F(\underline{Q})$. Thus the next optimal trajectory is given by

$$\begin{bmatrix} \underline{Y} \\ \overline{\underline{U}} \end{bmatrix}_n^* = \begin{bmatrix} \underline{Y} \\ \overline{\underline{U}} \end{bmatrix}_c^* + \alpha' P\underline{\delta} \tag{4.4}$$

where P is the projection operator (3.5) and

$$\alpha' = \frac{(\phi - 1)\underline{\delta}^T \tilde{Q}_c \left(\begin{bmatrix} \underline{Y} \\ \overline{\underline{U}} \end{bmatrix}_c^* - \begin{bmatrix} \underline{Y} \\ \overline{\underline{U}} \end{bmatrix}^d \right)}{(\phi - 1)\underline{\delta}^T H (H^T \tilde{Q}_c^{-1} H)^{-1} H^T \underline{\delta} - \phi \underline{\delta}^T \tilde{Q}_c \underline{\delta}} \tag{4.5}$$

Proof
The result follows immediately when (4.1) is substituted for $\underline{\gamma}$ in (3.10).

The correction term in (4.4), $\alpha' P\underline{\delta}$, is along the projection of δ on to $F(\underline{Q})$. This implies that the new optimal solution is as 'near' to the specified preferred soluton as allowed by the model which forms the whole or part of the manifold $F(\underline{b})$ defined by (3.3). The scalar α' has to be restricted and depends on the chosen ϕ and $\underline{\delta}$. The following theorem establishes the convergence of the method when the new optimal solutions are expressed via (4.4). This is done by proving, for restricted α', that the new optimal solution (4.4) is nearer to the preferred trajectory than the current optimal solution.

Theorem: Convergence (4.6)

If

$$\begin{bmatrix} \underline{Y} \\ \overline{\underline{U}} \end{bmatrix}_n^* = \begin{bmatrix} \underline{Y} \\ \overline{\underline{U}} \end{bmatrix}_c^* + \alpha' P\underline{\delta} \tag{4.4}$$

with $0 \leqslant \alpha' \leqslant 1$, then

$$\left\| \begin{bmatrix} \underline{Y} \\ \overline{\underline{U}} \end{bmatrix}^P - \begin{bmatrix} \underline{Y} \\ \overline{\underline{U}} \end{bmatrix}_c^* \right\| \geqslant \left\| \begin{bmatrix} \underline{Y} \\ \overline{\underline{U}} \end{bmatrix}^P - \begin{bmatrix} \underline{Y} \\ \overline{\underline{U}} \end{bmatrix}_n^* \right\| \tag{4.7}$$

Proof
Figure 6.3 gives a two-dimensional representation of what needs to be proved.

The vector $\underline{p}$ denotes, in short, the preferred trajectory (2.17), $\underline{c}$ denotes the current optimal trajectory (2.14), $\underline{n}$ denotes the new optimal trajectory (2.16) given by (4.4) and $\underline{q}$ denotes the projection of $\underline{p}$ on to $F(\underline{b})$ which is also equivalent to the new optimal trajectory (4.4) when $\alpha' = 1$. In terms of these vectors, the required results (4.7) becomes

$$\|\underline{\delta}\| \geqslant \|\underline{p} - \underline{n}\|. \tag{4.8}$$

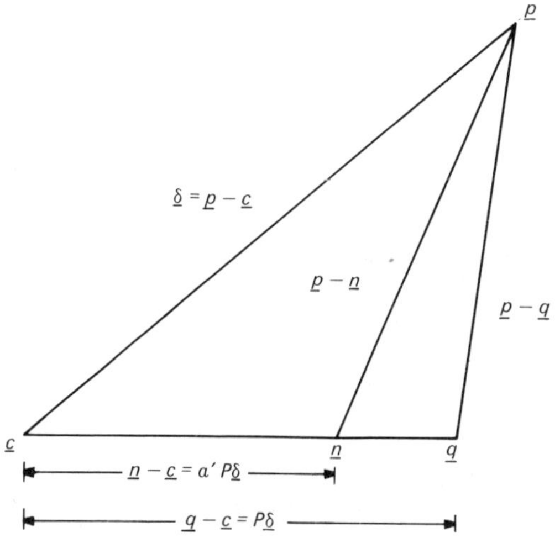

FIG. 6.3

Since $\underline{q}$ is the projection of $\underline{p}$, the inequality

$$< \underline{p} - \underline{q}, \underline{q} - \underline{c} > \geqslant 0 \tag{4.9}$$

holds (see, e.g., Luenberger (1969), pp. 69–70). Any vector $\underline{w}$ between $\underline{q}$ and $\underline{c}$, on $\underline{q} - \underline{c}$, may be written as a function of α

$$\underline{w}(\alpha) = \alpha \underline{q} + (1 - \alpha)\underline{c}, \; 0 \leqslant \alpha \leqslant 1. \tag{4.10}$$

Consider the function

$$f(\alpha) = < \underline{p} - \underline{w}(\alpha), \underline{w}(\alpha) - \underline{c} >$$

which, using (4.10) may be written as

$$\begin{aligned} f(\alpha) &= \alpha < \underline{p} - \underline{c}, \underline{q} - \underline{c} > - \alpha^2 < \underline{q} - \underline{c}, \underline{q} - \underline{c} > \\ &= \alpha < \underline{p} - \underline{q}, \underline{q} - \underline{c} > + \alpha < \underline{q} - \underline{c}, \underline{q} - \underline{c} > - \alpha^2 < \underline{q} - \underline{c}, \underline{q} - \underline{c} > \end{aligned}$$

Using (4.9)

$$f(\alpha) \geqslant 0, \; 0 \leqslant \alpha \leqslant 1. \tag{4.11}$$

is established. Furthermore,

$$\begin{aligned}||\underline{\delta}||^2 &= <\underline{\delta}, \underline{\delta}> \\ &= <(\underline{p}-\underline{n})+(\underline{n}-\underline{c}), (\underline{p}-\underline{n})+(\underline{n}-\underline{c})> \\ &= <\underline{p}-\underline{n}, \underline{p}-\underline{n}> + <\underline{p}-\underline{n}, \underline{n}-\underline{c}> + <\underline{n}-\underline{c}, \underline{n}-\underline{c}> \\ &\geqslant <\underline{p}-\underline{n}, \underline{p}-\underline{n}> + 2<\underline{p}-\underline{n}, \underline{n}-\underline{c}>\end{aligned}$$

and since

$$n = \underline{w}(\alpha') \text{ with } 0 \leqslant \alpha' \leqslant 1$$

using (4.11)

$$||\underline{\delta}||^2 \geqslant <\underline{p}-\underline{n}, \underline{p}-\underline{n}>$$

and thereby (4.8) is obtained.

5. NUMERICAL EXPERIENCE

The method for respecifying the weighting matrix of the quadratic objective function described in this paper was tested using the econometric model and the optimal control program discussed in Holly *et al.* (Chapter 8 of this volume). A particular feature of the dynamic programming algorithm for solving this optimal control problem is that the structure of the weighting matrix of the objective function (2.1) is restricted. This is discussed in detail in Section 5.1. To start the respecification procedure, the initial set of weights are taken to be those discussed in Holly *et al.* (Chapter 8 of this volume). The weighting matrix is respecified three times and the rationale behind each of these is given in Sections 5.2, 5.3, 5.4, and 5.5.

5.1 The Structure of the Objective Function

The objective function (2.1) used in the optimal control problem implicity restricts the type of weighting allowed on the deviations of $\underline{y}(k)$ and $\underline{u}(k)$ from their desired values. This restriction prevents the weighting of the deviations at a given time with deviations at any other time. Thus deviations at time k of all the variables (endogenous variables and policy instruments) may only be weighted against all deviations at the same time. The structure for $\grave{Q}$ that incorporates this constraint is the following:

$$\tilde{Q} \triangleq \left[\begin{array}{cccccc|cccccc}
A(1) & & & & & & C^T(1) & & & & & \\
 & A(2) & & & & & & C^T(2) & & & & \\
 & & \ddots & & & & & & \ddots & & & \\
 & & & A(k) & & & & & & C^T(k) & & \\
 & & & & \ddots & & & & & & \ddots & \\
 & & & & A(N-1) & & & & & & C^T(N-1) & \\
 & & & & & A(N) & & & & & & C^T(N) \\
\hline
C(1) & & & & & & B(1) & & & & & \\
 & C(2) & & & & & & B(2) & & & & \\
 & & \ddots & & & & & & \ddots & & & \\
 & & & C(k) & & & & & & B(k) & & \\
 & & & & \ddots & & & & & & \ddots & \\
 & & & & C(N-1) & & & & & & B(N-1) & \\
 & & & & & C(N) & & & & & & B(N)
\end{array}\right] \tag{5.1}$$

The submatrices $A(k)$, $B(k)$ and $C(k)$, $1 \leqslant k \leqslant N$, are the weights on the deviations of $\underline{y}(k)$, $\underline{u}(k)$ and $\underline{y}(k)$ against $\underline{u}(k)$, respectively.

To preserve the structure of (5.1) during an update (2.20), the preferred trajectory (2.17) is set equal to the current optimal trajectory (2.14) except at time k. At k, the modifications to the current optimal trajectory during this period are stored in the corresponding part of the preferred trajectory. Thus the displacement vector $\underline{\delta}$ (2.18) is zero except at k. If γ in (2.18) is set equal to zero except at time k, the structure of (5.1) is preserved during its respecification via (2.20) since the updating formula takes the form

$$\begin{bmatrix} A(k) & C^T(k) \\ C(k) & B(k) \end{bmatrix}_n = \begin{bmatrix} A(k) & C^T(k) \\ C(k) & B(k) \end{bmatrix}_c + \frac{\left(\underline{\gamma}_k - \begin{bmatrix} A(k) & C^T(k) \\ C(k) & B(k) \end{bmatrix}_c \underline{\delta}_k\right)\left(\underline{\gamma}_k - \begin{bmatrix} A(k) & C^T(k) \\ C(k) & B(k) \end{bmatrix}_c \underline{\delta}_k\right)^T}{\left\langle \underline{\gamma}_k - \begin{bmatrix} A(k) & C^T(k) \\ C(k) & B(k) \end{bmatrix}_c \underline{\delta}_k, \underline{\delta}_k \right\rangle} \tag{5.2}$$

where $\underline{\gamma}_k$ and $\underline{\delta}_k$ are the respective parts of $\underline{\gamma}$ and $\underline{\delta}$ corresponding to the preferred trajectory at k. The new weighting matrix $\tilde{Q}_n$ may thus be written as

$$\tilde{Q}_n = \left[\begin{array}{cc:ccc|cc:ccc}
A_c(1) & & & & & C_c^T(1) & & & & \\
& A_c(2) & & & & & C_c^T(2) & & & \\
\hdashline
& & A_c(k)+\Delta A(k) & & & & & C_c^T(k)+\Delta C^T(k) & & \\
& & & \ddots & & & & & \ddots & \\
& & & A_c(N-1) & & & & & C_c^T(N-1) & \\
& & & & A_c(N) & & & & & C_c^T(N) \\
\hline
C_c(1) & & & & & B_c(1) & & & & \\
& C_c(2) & & & & & B_c(2) & & & \\
\hdashline
& & C_c(k)+\Delta C(k) & & & & & B_c(k)+\Delta B(k) & & \\
& & & \ddots & & & & & \ddots & \\
& & & C_c(N-1) & & & & & B_c(N-1) & \\
& & & & C_c(N) & & & & & B_c(N)
\end{array}\right] \tag{5.3}$$

where $\Delta A(k)$, $\Delta B(k)$ and $\Delta C(k)$ are the respective corrections added to $A_c(k)$, $B_c(k)$ and $C_c(k)$ in (5.2). Thus (5.3) has the same structure as (5.1) and only those positions corresponding to time k are altered.

5.2 Some Numerical Results

The results reported in this section are only meant to convey the flavour of the method. They do not amount to an exhaustive attempt to produce a completely acceptable vector of trajectories since without the active participation of a policy-maker who formulates policies in the real world a 'completely acceptable' set cannot be defined.

The cited example is based upon the optimal control runs provided in Holly *et al.* (Chapter 8 of this volume). In particular, the respecification begins with the trajectories in Case II where the exogenous variable paths, $\underline{e}(k)$, $k = 1, 20$, were set equal to their historical values between 1970(1) and 1974(4).

A major weakness of the use of objective functions, quadratic or otherwise, in the formulation of economic policy is that the policy-maker is then required to attach 'cardinal' measures to marginal changes in his objectives and to take into account at the same time the numerous trade-offs between 'a little bit more of one thing and a little bit less of another'. Even the simplest of questions about his true objectives and their ordering will be sufficient to defeat him. With the method presented in this paper and illustrated below, the policy-maker is only asked to express reservations about the trajectories (if he has any) and to indicate a more preferred state of affairs. The rest follows automatically within the overall optimisation procedure.

5.3 First Respecification

The initial choice of weights in Holly *et al.* Table 8.1, p. 167 below) was made in a partly *ad hoc* manner. The weights on the tax controls were chosen deliberately so as to rule out their use in any other than the second quarter of each year and u_2, u_4 and u_5 were screened out along with all but y_1, y_{11}, y_{12} and y_{18} among the endogenous variables. There appeared no obvious way, however, of deciding what relative weights to use for the targets and instruments that remained, or what sort of time profile the weights on individual variables should have. Hence the targets were weighted equally with a weight of 2 throughout the entire twenty periods. For the controls, the main burden was designed to fall upon the exchange rate (u_3) and the tax controls (u_6, u_7) in the second quarter of each year. Some greater preference was given to government current expenditure (u_1) and national insurance contributions (u_8) tracking their desired trajectories by assigning to them a larger weighting relative to the other three effective controls. Nevertheless, because of the considerable impact that government expenditure can have on the behaviour of the model it was still employed very extensively. On the other hand u_8 was not used at all because u_6 and u_7 were able to bring about the necessary changes in disposable income despite the constraints on the frequency with which they could be used. Thus already we are learning something about the ways in which the structure of the model and the choice of weights influence the orchestration of instruments and the achievement of targets.

The trajectories of the targets and instruments after the run of the optimal control program for Case II are listed in Table 6.2 below. A number of undesirable characteristics (apart from those which reflect the particular structure of the model as discussed in Chapter 8) can easily be identified. Consider period 1972(2), for example. Aggregate allowances are reduced by 4 per cent. It is difficult to imagine a Chancellor of the Exchequer actually reducing the nominal value of personal allowances. This then is an ideal candidate for respecification. During 1972(2) imports (y_{12}) rise by 2.12 per cent, and prices (y_{18}) rise by 1.59 per cent (quarter on quarter). Both these rises are on the high side and so they are included in the respecification also.

The preferred values are then specified as given in Table 6.1 along with

TABLE 6.1 Details of the First Update

VARIABLES AT	PREFERRED VALUES:	CURRENT OPTIMAL VALUES:	$\underline{\delta}$
(1972(2))	Only those different from the current optimal values are listed, the rest are set equal to their current optimal value	declared undesirable by the policy-maker; at (1972(2))	(only the non-zero elements are listed)
y_{12}	1.5	2.1216	−.6216
y_{18}	1.2	1.5895	−.3895
u_7	0.0	−4.0036	4.0036

WEIGHTS

Current Weights at 1972(2), for other quarters see Holly et al. (Table 8.1 below).

A: endogenous variables:
$a_{1,1} = 2.0$, $a_{11,11} = 2.0$, $a_{12,12} = 2.0$, $a_{18,18} = 2.0$
B: policy instruments:
$b_{1,1} = 1.0$, $b_{2,2} = 3.0$, $b_{3,3} = .01$, $b_{4,4} = 3.0$, $b_{5,5} = 3.0$
$b_{6,6} = .01$, $b_{7,7} = .01$, $b_{8,8} = 1.0$

All other elements of matrices A and B and the whole of matrix C are set to zero.

New Weights (with $\gamma = \phi \grave{Q}_c \underline{\delta}$, $\phi = 3$)

A: endogenous variables:
$a_{1,1} = 2.0$, $a_{11,11} = 2.0$, $a_{12,12} = 4.4\text{–}98$, $a_{18,18} = 2.9817$,
$a_{18,12} = a_{12,18} = 1.5665$
B: policy instruments:
$b_{1,1} = 1.0$, $b_{2,2} = 3.0$, $b_{3,3} = .01$, $b_{4,4} = 3.0$, $b_{5,5} = 3.0$
$b_{6,6} = .01$, $b_{7,7} = .0126$, $b_{8,8} = 1.0$
C: y–u weights:
$c_{7,12} = -.0805$ $c_{7,18} = -.0505$

the corresponding values for $\underline{\delta}$ and the respecified weighting matrices.

We ask that y_{12} move more towards its original desired value, that y_{18} actually go below its original desired value, while u_7 is required to move towards zero. The objective function was then updated and the optimal control program was rerun. The results showing the new trajectories are listed in Table 6.2. Some improvement has taken place. While u_7 seems to move very readily towards its preferred value, the values for both y_{12} and y_{18} hardly change at all. This is clearly due to the relative weightings of policy instruments being higher than those for the endogenous variables. Since u_7 has changed it might be wondered where this instrument's impact makes itself felt. The closest analogue of u_7, given the form of the cost function, is u_6, the basic tax rate. The original and new trajectories for u_6 are given in Table 6.2. The basic tax rate, to compensate for the constraint upon u_7 going negative, has to rise by 8.58 per cent rather than 5.30 per cent.

5.4 Second Respecification

The optimal control law requires that government current expenditure (u_1) increase by 4.63 per cent in the first quarter of the control period (after the first respecification). As compared with the historical behaviour of government expenditure this represents a very rapid and probably infeasible increase during one single quarter. It was therefore decided that the preferred value for u_1 should be -1.0 per cent (see Table 6.3). This was not necessarily because we actually wanted u_1 to be negative during 1970(1). However, the first respecification has indicated that one way of pulling the trajectory towards a desired region is to overstress the direction in which one wants the trajectory to go. The result of doing this is shown in Table 6.4 for all the targets and instruments. The trajectories here should be compared to the new value columns of the previous table. Since government current expenditure is such an important control we would expect that it would have a major impact upon the configuration of targets and instruments. The rate of increase of u_1 in period 1970(1) is reduced to 2.58 per cent, and the reduction is compensated for by part of the increased government expenditure being shifted to the second quarter. More interestingly, some of the work formerly done by u_1 is taken on by u_6 and u_7. Basic tax rate is cut by 11.42 per cent rather than 6.29 per cent; while allowances are increased by 8.63 per cent as compared with 4.75 per cent previously.

The most interesting consequence, however, of this simple respecification is for unemployment. The increase in unemployment is not

TABLE 6.2 Original (Current) Optimal Trajectories and the Results of the first respecification (1972(2)), $\phi = 3$. All variables in $\%\Delta$.

ENDOGENOUS VARIABLES

	y_1		y_{11}		y_{12}		y_{18}	
	Current Optimal Trajectory	*New Optimal Trajectory*	*Current Optimal Trajectory*	*New Optimal Trajectory*	*Current Optimal Trajectory*	*New Optimal Trajectory*	*Current Optimal Trajectory*	*New Optimal Trajectory*
70(1)	2.7068	2.7087	−.0661	−.0661	2.4947	2.4941	1.2606	1.2610
70(2)	−.0582	−.0573	1.9629	1.9629	2.3969	2.3987	.0248	.0245
70(3)	.1760	.1770	−.7611	−.7611	.6272	.6284	1.1677	1.1682
70(4)	−.6972	−.6981	2.9965	2.9965	1.5588	1.5604	.5793	.5769
71(1)	−.5907	−.5900	.8643	.8502	1.2230	1.2277	.8889	.8853
71(2)	.0564	.0589	1.5026	1.4939	1.2381	1.2395	.9438	.9305
71(3)	.4824	.5067	.9104	.8669	1.3397	1.3375	.9728	.9628
71(4)	−.1045	−.1196	.9240	.9342	1.7308	1.7523	.7377	.7045
72(1)	.2694	.2215	1.1065	1.0839	.7829	.7973	1.1354	1.0989
72(2)	−.3146	−.2682	1.3569	1.3235	2.1216	2.0637	1.5895	1.5770
72(3)	.4684	.5318	1.0025	1.0163	.5718	.5751	1.0020	.9839
72(4)	−.0881	−.0843	1.1362	1.2569	2.2144	2.2366	.8757	.8521
73(1)	.0629	.0506	.3390	.3427	1.0859	1.0840	.5586	.5551
73(2)	.3559	.3384	1.7527	1.7704	1.4210	1.4212	1.1029	1.1018
73(3)	.1975	.1483	.9236	.9136	1.4084	1.4082	.9794	.9804
73(4)	.1674	.1692	2.1642	2.1367	1.2204	1.2162	1.3497	1.3511
74(1)	−.4679	−.4648	.8369	.8418	1.2509	1.2528	1.4129	1.4095
74(2)	.2616	.2624	1.7305	1.7305	1.4837	1.4851	2.6847	2.6805
74(3)	.5457	.5460	1.0779	1.0791	1.3223	1.3215	1.8600	1.8566
74(4)	−.4549	−.4550	1.1274	1.1283	1.7764	1.7763	1.9329	1.9321

TABLE 6.2 (continued)

POLICY INSTRUMENTS

	u_1		u_3		u_6		u_7	
	Current Optimal Trajectory	*New Optimal Trajectory*	*Current Optimal Trajectory*	*New Optimal Trajectory*	*Current Optimal Trajectory*	*New Optimal Trajectory*	*Current Optimal Trajectory*	*New Optimal Trajectory*
70(1)	4.6358	4.6321	−1.9200	−1.8902	−.0252	−.0253	.0191	.0191
70(2)	1.6295	1.6253	.4852	.5035	−6.2487	−6.2935	4.7199	4.7538
70(3)	.8440	.8356	−1.9800	−1.8890	−.0027	−.0026	.0020	.0020
70(4)	−.5592	−.5613	.6229	.6016	.0116	.0115	−.0087	−.0087
71(1)	.1766	.1605	−.7763	−.7290	.0065	.0067	−.0049	−.0051
71(2)	.9954	.9933	.6162	.6860	.0241	.0092	−.0182	−.0070
71(3)	.7052	.6828	−3.2890	−3.3179	.0030	.0032	−.0023	−.0024
71(4)	.6584	.7077	5.6179	5.3653	.0041	.0035	−.0031	−.0026
72(1)	.9035	.9227	−.5773	−.5859	.0020	.0017	−.0015	−.0013
72(2)	1.0093	.9612	−.2146	−.2516	5.3003	8.5786	−4.0036	−.5925
72(3)	1.2801	1.3119	−2.8207	−2.7998	−.0010	−.0018	.0008	.0013
72(4)	.5428	.4969	1.1611	1.2186	.0076	.0075	−.0058	−.0057
73(1)	−.0340	−.0700	−1.0318	−1.0421	.0124	.0122	−.0094	−.0092
73(2)	1.1504	1.1325	.3652	.3651	−1.2805	−1.2810	.9672	.9676
73(3)	.6394	.6348	−2.6745	−2.6770	.0031	.0031	−.0024	−.0024
73(4)	.6341	.6430	.3265	.3246	.0018	.0017	−.0014	−.0013
74(1)	.7597	.7623	9.8481	9.8298	0	−.0001	0	0
74(2)	1.2943	1.2939	−12.5547	−12.5605	−7.9951	−7.9912	6.0391	6.0361
74(3)	.5402	.5399	0	0	−.0009	−.0009	.0007	.0007
74(4)	.0288	.0285	0	0	.0036	.0036	−.0027	−.0027

TABLE 6.3 Details of the Second Respecification

VARIABLE AT	PREFERRED VALUES	CURRENT OPTIMAL VALUES	$\underline{\delta}$
(1970(1))	Only those different from the current optimal values are listed, the rest are set equal to their current optimal value	declared undesirable by the policy maker; at (1970(1))	(Only the non-zero elements are listed)
u_1	−1.0	4.6321	−5.6321

WEIGHTS

Current Weights at 1970(1), for other quarters see the new weights in Table 6.1 and Holly et al. (Table 8.1).

A: endogenous variables:
$a_{1,1} = 2.0$ $a_{11,11} = 2.0$ $a_{12,12} = 2.0$ $a_{18,18} = 2.0$
B: policy instruments:
$b_{1,1} = 1.0$ $b_{2,2} = 3.0$ $b_{3,3} = .01$ $b_{4,4} = 3.0$
$b_{5,5} = 3.0$ $b_{6,6} = 15.0$ $b_{7,7} = 15.0$ $b_{8,8} = 1.0$

All other elements of A, B and the whole of matrix C are zero.

New Weights (with $\underline{\gamma} = \phi \tilde{Q}_c \underline{\delta}$, $\phi = 3$)

A: endogenous variables:
Same as the current weights.
B: policy instruments:
$b_{1,1} = 3.0$; the rest are the same as the current weights.
C: $y - u$ weights:
Null matrix.

checked until the fourth quarter though by the second year the shape of the trajectory is very similar to that previously obtained. The trajectories of the remaining targets are hardly altered at all.

5.5 Third Respecification

For this respecification we return to 1972(2), the subject of the first respecification. Though u_7 was reduced markedly it still has a negative value. Equally, y_{12} and y_{18} did not move very far towards the preferred values. The policy-maker might insist that u_7 be positive and set the new desired value at 3.5 per cent. At the same time he changes his mind about

TABLE 6.4 New optimal trajectories obtained via the second respecification 1970(1), $\phi = 3$. All variables in % Δ.

	ENDOGENOUS VARIABLES				POLICY INSTRUMENTS			
	y_1	y_{11}	y_{12}	y_{18}	u_1	u_3	u_6	u_7
70(1)	3.7145	−.0661	2.1421	1.5018	2.5827	−1.6984	−.0413	.0312
70(2)	.5995	1.9629	2.7211	−.0794	2.3145	.6417	−11.4206	8.6265
70(3)	.3254	−.7611	.7579	1.1690	.9257	−1.7995	−.0025	.0019
70(4)	−.6913	2.9965	1.4761	.5990	−.5565	.6733	.0116	−.0088
71(1)	−.6158	.7585	1.2124	.9531	.1536	−.6961	.0070	−.0053
71(2)	.0603	1.4278	1.2060	.9762	1.0003	.7008	.0976	−.0737
71(3)	.5024	.8241	1.3457	1.0076	.6714	−3.3088	.0035	−.0026
71(4)	−.1177	.8999	1.7409	.7182	.6933	5.3792	.0037	−.0028
72(1)	.2136	1.0682	.8012	1.1019	.9053	−.6017	.0020	−.0015
72(2)	−.2690	1.3164	2.0488	1.5758	.9530	−.2846	8.8568	−.8590
72(3)	.5267	1.0119	.5777	.9767	1.2958	−2.8351	−.0014	.0011
72(4)	−.0882	1.2502	2.2346	.8434	.4802	1.1700	.0079	−.0060
73(1)	.0475	.3507	1.0812	.5443	−.0768	−1.0407	.0125	−.0094
73(2)	.3382	1.7862	1.4173	1.0935	1.1217	.3653	−1.2773	.9648
73(3)	.1846	.9305	1.4067	.9777	.6258	−2.6730	.0032	−.0024
73(4)	.1646	2.1600	1.2164	1.3512	.6360	.3277	.0018	−.0013
74(1)	−.4672	.8412	1.2499	1.4135	.7628	9.8436	−.0001	0
74(2)	.2629	1.7305	1.4831	2.6837	1.2955	−12.5656	−8.0121	6.0519
74(3)	.5465	1.0772	1.3213	1.8591	.5399	0	−.0009	.0007
74(4)	−.4558	1.1269	1.7756	1.9328	.0280	0	.0036	−.0027

TABLE 6.5 Details of the Third Respecification

VARIABLES AT	PREFERRED VALUES:	CURRENT OPTIMAL VALUES:	$\underline{\delta}$
(1972(2))	Only those different from the current optimal values are listed, the rest are set equal to their current optimal value	declared undesirable by the policy-maker; at (1972(2))	(only the non-zero elements are listed)
y_{12}	1.5	2.0488	−.5488
y_{18}	1.0	1.5758	−.5758
u_1	.3	.9530	−.6530
u_7	3.5	−.8590	4.359

WEIGHTS

Current Weights at 1972(2) are the new weights computed in Table 6.1. For other quarters see Table 6.3 and Holly et al. (Table 8.1).

New Weights (with $\underline{\gamma} = \phi \breve{Q}_c \underline{\delta}$)
$\phi = 3$

A: endogenous variables:
$a_{1,1} = 2.0$, $a_{11,11} = 2.0$, $a_{12,12} = 10.4750$, $a_{18,12} = a_{12,18} = 6.0555$
$a_{18,18} = 6.3541$
B: policy instruments:
$b_{1,1} = 1.1839$ $b_{2,2} = 3.0$ $b_{3,3} = .01$ $b_{4,4} = 3.0$ $b_{5,5} = 3.0$ $b_{6,6} = .01$
$b_{7,7} = .0197$ $b_{1,7} = b_{7,1} = -.0361$ $b_{8,8} = 1.0$
C: $y-u$ weights:
$c_{1,12} = 1.0481$, $c_{1,18} = .7874$, $c_{7,12} = -.2862$, $c_{7,18} = -.2050$
$\phi = 10$

A: endogenous variables:
$a_{1,1} = 2.0$ $a_{11,11} = 2.0$ $a_{12,12} = 31.3885$ $a_{18,12} = a_{12,18} = 21.7669$
$a_{18,18} = 18.1573$
B: policy instruments:
$b_{1,1} = 1.8274$, $b_{2,2} = 3.0$, $b_{3,3} = .01$, $b_{4,4} = 3.0$, $b_{5,5} = 3.0$
$b_{6,6} = .01$, $b_{7,7} = .0445$, $b_{7,1} = b_{1,7} = -.1623$, $b_{8,8} = 1.0$
C: $y-u$ weights:
$c_{1,12} = 4.7167$, $c_{1,18} = 3.5434$, $c_{7,12} = -1.0059$, $c_{7,18} = -.7457$

All non-specified elements of A, B and C are set to zero.

TABLE 6.6 New optimal trajectories obtained via the third respecification (1972(2)), $\phi = 3$. All variables in $\%\Delta$.

	ENDOGENOUS VARIABLES				POLICY INSTRUMENTS			
	y_1	y_{11}	y_{12}	y_{18}	u_1	u_3	u_6	u_7
70(1)	3.7159	−.0661	2.1416	1.5022	2.5798	−1.6253	−.0412	−.0311
70(2)	.6052	1.9629	2.7194	−.0784	2.3085	.7011	−11.4123	8.6203
70(3)	.3318	−.7611	.7612	1.1691	.9091	−1.6286	−.0024	.0018
70(4)	−.6912	2.9965	1.4821	.5914	−.5668	.6972	.0118	−.0089
71(1)	−.6158	.7236	1.2227	9398	.1305	−.6128	.0074	−.0056
71(2)	.0682	1.3994	1.2057	.9449	1.0028	.7329	.2329	−.1759
71(3)	.5391	.7425	1.3486	.9646	.6455	−3.3332	.0039	−.0029
71(4)	−.1345	.8885	1.7727	.6429	.7705	5.0625	.0030	−.0023
72(1)	.1228	1.0283	.8361	1.0155	.9721	−.6179	.0015	−.0011
72(2)	−.1896	1.3011	2.0073	1.5383	.4777	−.3496	11.9667	4.9242
72(3)	.6331	1.0236	.5856	.9319	1.3772	−2.8192	−.0027	.0021
72(4)	−.0703	1.4016	2.2680	.7974	.4312	1.2338	.0077	−.0058
73(1)	.0365	.3584	1.0790	.5307	−.1249	−1.0562	.0122	−.0092
73(2)	.3156	1.8172	1.4174	1.0874	1.0910	.3646	−1.2762	.9640
73(3)	.1659	.9229	1.4061	.9773	.6115	−2.6769	.0033	−.0025
73(4)	.1646	2.1295	1.2109	1.3524	.6413	.3252	.0017	−.0013
74(1)	−.4650	.8486	1.2520	1.4091	.7630	9.8178	.0001	0
74(2)	.2633	1.7308	1.4842	2.6786	1.2939	−12.5742	−7.9975	6.0409
74(3)	.5467	1.0790	1.3204	1.8552	.5300	0	−.0009	.0007
74(4)	−.4562	1.1280	1.7755	1.9323	.0276	0	.0036	−.0027

TABLE 6.7 New optimal trajectories obtained via the third respecification (1972 (2)), $\phi = 10$. All variables in %Δ.

	ENDOGENOUS VARIABLES				POLICY INSTRUMENTS			
	y_1	y_{11}	y_{12}	y_{18}	u_1	u_3	u_6	u_7
70(1)	3.7171	−.0661	2.1411	1.5025	2.5775	−1.5645	−.0412	−.0311
70(2)	.6099	1.9629	2.7180	−.0776	2.3036	.7506	−11.4055	8.6151
70(3)	.3372	−.7611	.7639	1.1692	.8953	−1.4865	−.0022	.0017
70(4)	−.6910	2.9965	1.4872	.5851	.5754	.7171	.0119	−.0090
71(1)	−.6159	.6946	1.2313	.9287	.1113	−.5435	.0077	−.0058
71(2)	.0748	1.3758	1.2055	.9189	1.0048	.7596	.3455	−.2609
71(3)	.5695	.6746	1.3510	.9331	.6239	−3.3534	.0042	−.0032
71(4)	−.1485	.8790	1.7992	.5803	.8346	4.7991	.0024	−.0018
72(1)	.0473	.9952	.8651	.9436	1.0277	−.6313	.0011	−.0008
72(2)	−.1236	1.2823	1.9727	1.5070	.0825	−.4036	14.5522	9.7322
72(3)	.7216	1.0332	.5921	.8947	1.4449	−2.8060	−.0038	.0029
72(4)	−.0553	1.5275	2.2957	.7592	.3905	1.2868	.0075	−.0057
73(1)	.0274	.3648	1.0772	.5194	−.1649	−1.0691	.0120	−.0090
73(2)	.2969	1.8430	1.4176	1.0824	1.0655	.3640	−1.2753	.9633
73(3)	.1503	.9165	1.4056	.9770	.5996	−2.6801	.0034	−.0025
73(4)	.1647	2.1041	1.2062	1.3533	.6458	.3231	.0017	−.0012
74(1)	−.4632	.8547	1.2538	1.4055	.7632	9.7963	−.0001	0
74(2)	.2637	1.7311	1.4851	2.6745	1.2926	−12.5813	−7.9854	6.0317
74(3)	.5468	1.0806	1.3196	1.8519	.5381	0	−.0009	.0007
74(4)	−.4565	1.1290	1.7753	1.9319	.0273	0	.0036	−.0028

the preferred value of y_{18} and reduces it from 1.2 to 1.0 per cent. Though capricious, the policy-maker is not irrational. His preferred value for y_{12} remains at 1.5 per cent. The policy maker also decides that in 1970(2) he wants government current expenditure to increase by 0.3 per cent rather than .95 (see Table 6.5). Again it is not necessary that he justifies his choice. The setting of the value of ϕ can be used in this context to put additional pressure on y_{12} and y_{18} to move towards their preferred values. Two examples are given: the first with $\phi = 3$ and the second with $\phi = 10$. The results of the first respecification ($\phi = 3$) is given in Table 6.6, where this is compared with the result of the second respecification. Again the values for y_{12} and y_{18} are little changed.

The effect of the third respecification on u_7 is once more marked. From a value of $-.86$ it increases to 4.92. To compensate for this u_6 goes from 8.86 to 12.00 per cent. Clearly the policy maker has to decide the trade-off between increases in allowances and increases in basic tax rate and this he can do by using the respecification method. The increase in government current expenditure (u_1) in 1972(2) is reduced from .95 to .48 but as with the second respecification this is compensated for by changes in government current expenditure in neighbouring quarters.

When $\phi = 10$ (Table 6.7) we are more successful in forcing y_{12}, y_{18} towards the preferred values. However, both u_7 and u_1 overshoot their preferred trajectories with consequent offsetting changes in u_6 and in u_1 in the neighbourhood of 1972(2). Since this overshooting was not objectionable the respecification was terminated. If it were desired to rectify this overshooting, the same run could be repeated with $\phi = 10$ and lower preferred values for u_1, u_7 and the same preferred values for y_{12} and y_{18}.

The iterative respecification need not stop here. The trajectory for the exchange rate has a number of possibly undesirable elements particularly in 1974 and 1971. Applying the procedures described above will enable the policy maker to explore the consequences of modifying this exchange rate trajectory, or any other. It will also tell him much about the nature of the economic model and the kinds of hard economic constraints that face him when decisions have to be made.

6. CONCLUSIONS

The difficulty about the specification of the correct weighting matrix of a quadratic objective function is the scarcity of information about it. The method described in this paper treats this problem at two levels of information. The first is when gradient information about the true objective function is available. In this case Theorem (3.1) establishes convergence to the true weights of the quadratic function. At the second level this information is not available and only the preferred trajectories are specified. Updating the weighting matrix at this level is aimed only at producing a new optimal trajectory which is more acceptable to the policy maker than the current one (see Theorem (4.6)). In both cases the effect of the method in the modification of the optimal trajectories has been investigated.

The numerical results discussed in Section 5 are only expected to illustrate the method.

ACKNOWLEDGEMENT

The work described here was carried out as part of the Programme of Research into Econometric Methods. The financial support of the Social Science Research Council is gratefully acknowledged. The authors also thank Dr K. Velupillai of Lund University for many helpful discussions.

NOTES

1. Other alternatives to generating optimal solutions of (2.9) satisfying Ω are either respecifying the desired trajectories or imposing additional inequality constraints (Livesey (1971, 1976)) on the values that the solution of (2.9) may take. The former option is assumed to be not allowed since the desired values have been fixed by the policy-maker.

   The method described in this paper is conceptually equivalent to imposing inequality constraints as discussed in Rustem *et al.* (1976). The method of this paper may be superior to the latter alternative, when imposing inequality constraints is not computationally feasible. The iterative procedure described in Section 3 also permits systematic corrections to subsequent optimal solutions of (2.9) by identifying the trade-offs between variables. This helps to expand the knowledge of the policy-maker as to what can be required from the economic system at that particular time.

REFERENCES

Bray, J. (1972) 'Some Considerations in the Practice of Economic Management', PREM Discussion Paper no. 2.

Bray, J. (1975), 'Optimal Control of a Noisy Economy with the U.K. as an Example', *J. R. Statist. Soc.*, series A, vol. 138, part 3.

Broyden, C. (1967), 'Quasi-Newton Methods and their Application to Function Minimisation', *Math. of Comp.*, v. 21, pp. 368–81.

Canon, M. D., Cullum, C. D., and Polak, E. (1970), *Theory of Optimal Control and Mathematical Programming*, (McGraw-Hill, New York).

Davidon, W. C. (1968) 'Variance Algorithm for Minimization', *Comp. J.*, v. 10, pp. 406–10.

Fiacco, A.V., and McCormick, G. P. (1968), *Nonlinear Programming: Sequential Unconstrained Minimization Techniques* (John Wiley, New York).

Friedman, B. (1972), 'Optimal Economic Stabilization Policy: An Extended Framework', *J. Polit. Econ.* vol. LXXX, pp. 1002–22.

Frisch, R. (1972), 'Cooperation between Politicians and Econometricians on the Formalization of Political Preferences', University of Oslo, Institute of Economics Reprint Series No. 90.

Holly, S. *et al.* (1977), 'Control Exercises with a Small Linear Model of the U.K. Economy' (Chapter 8 of this volume).

Livesey, D. (1973), 'Can Macro-Economic Planning Problems Ever be Treated as a Quadratic Regulator Problem? *IEE Conference Publications*, vol. 101, pp. 1–14.

Livesey, D. (1976), 'Feasible Directions in Economic Policy', presented at the European Meeting of the Econometric Society, Helsinki, 1976.

Luenberger, D. (1969), *Optimization by Vector Space Methods* (John Wiley, New York).

Murtagh, B. A., and Sargent, R. W. H. (1969), 'A Constrained Minimization Method with Quadratic Convergence', in R. Fletcher (ed.), *Optimization* (Academic Press, New York and London).

Murtagh, B. A., and Sargent, R. W. H. (1970), 'Computational Experience with Quadratically Convergent Minimization Methods', *Comp. J.*, vol. 13, pp. 185–94.

Polak, E. (1971), *Computational Methods in Optimization* (Academic Press, New York and London).

Rustem, B. (1978), 'Projection Methods in Constrained Optimization', PhD dissertation University of London.

Rustem, B., Velupillai, K., and Westcott, J. H. (1976), 'A Method for Respecifying the Weighting Matrix of a Quadratic Criterion Function', PREM Discussion Paper no. 15.

Westcott, J. H., Holly, S., Rustem, B., and Zarrop, M. (1976), 'Control Theory in the Formulation of Economic Policy', Submission to the Committee of Inquiry on Policy Optimisation.

7 The Sensitivity of Optimal Policies to Stochastic and Parametric Changes

A. Hughes-Hallett

I. INTRODUCTION

Traditionally econometric models are used to find acceptable policy strategies by extensive simulation exercises. To resolve conflicts between rival target variables or between rival values of a target over time, or to place costs on large variations in instrument variables, some index of overall performance must be employed — but there is no guarantee we approach the best policies. The success of optimal control methods in other contexts therefore makes it very tempting to apply them for policies to optimise that index. Since recent results indicate that, computing constraints apart, the mathematical and dimension difficulties of the general problem are now largely solved, it is now important to gain experience of the specification of the economic aspects of the problem. Virtually no experimentation of this kind has been undertaken.

Policy-makers are often suspicious of optimal control methods. It is easy to reject the technique outright because one does not trust the model in its structure, the reliability of its estimates, its ability to forecast accurately, or the accuracy of its exogenous information, or because one cannot precisely specify the economic priorities, and desired positions, in the form of an objective function. It is necessary therefore to see if indeed the policies are sensitive to any of these problems and, if so, to which item so that we can concentrate on its specification. Methods are given here to compute alterations to the policies due either to changes of a given size in any of the parameters or to random shocks to the problem. Since the results are optimal conditional on the objective function and model information, we can also regard their alterations as indicating the consequences of changing the economic specification of our problem. The only alternative procedure is to run the control algorithm for each case—i.e. Monte Carlo simulation over the conditioning information which are tedious computationally and completely problem dependent. The methods here can be applied to any problem of its kind.

II. THE PROBLEM DEFINITION

An estimated linear econometric model is available to the planner in structural form

$$\beta Y_t + \Gamma \overline{X}_t = U_t \tag{1}$$

The restricted reduced form is, with maximum lag in endogenous variables ℓ and in instruments p,

$$Y_t^* = \Pi_1 Y_{t-1}^* + \Pi_2 X_t^* + \begin{pmatrix} \Pi_3 \\ 0 \end{pmatrix} N_t + \begin{pmatrix} \beta^{-1} \\ 0 \end{pmatrix} U_t \tag{2}$$

We partition $\overline{X}_t = [Y_{t-1}^{*'} \vdots X_t^{*'} \vdots N_t']'$ together with $X_t^* = [X_t' \ldots . X_{t-p}']'$,

$$\Pi_1 = \begin{bmatrix} \Pi_{11} \ldots . & \Pi_{1\ell} \\ I \quad \ldots . & 0 \\ \quad . & \vdots \\ 0 \quad . I & 0 \end{bmatrix}, \Pi_2 = \begin{bmatrix} \Pi_{20} \ldots . \Pi_{2p} \\ 0 \end{bmatrix},$$

$$Y_t^* = [Y_t' \ldots Y_{t-\ell+1}']'$$

X_t^* is an n^*-vector of potential policy instruments available to the planner (government), N_t is an r-vector of noncontrollable (strictly exogenous to the planner) variables—and any of their lags are treated as separate variables in N_t. Y_t^* is an m^*-vector of potential targets of the policy makers. Let the planning horizon be $t = 1, \ldots, T$ periods. All variables are assumed bounded above and below throughout the planning and sample periods. U_t follows some unspecified probability density function with finite mean and variance. Until Section V we ignore the stochastic nature of β, Γ and the possibility of it in N_t. The t-period final forms[1] are stacked up for $t = 1, \ldots ., T$ so that

$$Y = RX + s \tag{3}$$

where $Y = [Y'_1 \ldots Y'_T]'$; $X = [X'_1 \ldots X'_T]'$; $s = \sigma + SU$ with $\sigma' = [\sigma'_1 \ldots \sigma'_T]$ and $u' = [u'_1 \ldots u'_T]$;

$$R = \begin{bmatrix} R_1 & & 0 \\ \cdot & \cdot & \\ \cdot & & \cdot \\ \cdot & & \cdot \\ R_T & \ldots & R_1 \end{bmatrix} \quad \text{with } R_i = \sum_{j=1}^{\min(p,i)} \Pi_1^{i-1} \begin{pmatrix} \Pi_{2,j-1} \\ 0 \end{pmatrix}$$

$$S = \begin{bmatrix} S_1 & & 0 \\ \cdot & \cdot & \\ \cdot & & \cdot \\ \cdot & & \cdot \\ S_T & \ldots & S_1 \end{bmatrix} \text{with } S_i = \Pi_1^{i-1}\beta^{-1}$$

and

$$\sigma_i = \Pi_1 \sigma_{i-1} + \Pi_3 N_i + \sum_{j=i}^{p} \begin{pmatrix} \Pi_{2j} \\ 0 \end{pmatrix} X_{i-j}$$

where

$$\sigma_1 = \Pi_1 Y^*_{t=0} + \Pi_3 N_1 + \sum_{j=1}^{p} \begin{pmatrix} \pi_{2j} \\ 0 \end{pmatrix} X_{1-j}.$$

For any particular planning exercise a subset of m targets and n instruments will have been selected after due deliberation by the planning authorities. The m^*-m non-targets are therefore assumed discarded from (3) and the n^*-n non-instruments transferred to N_t. R is thus mT by nT.

Desired trajectories of targets and instruments Y^d_t, X^d_t, $t = 1, \ldots, T$ are selected after deliberation by the planning authorities. Further, the authorities' preferences between rival targets, rival instruments, and their trade-offs for variables taking conflicting values over time, are specified in a quadratic preference function (welfare function)

$$w = \tfrac{1}{2}\tilde{Z}'Q\tilde{Z} + q'\tilde{Z} \tag{4}$$

reflecting their economic priorities. $\tilde{Z} = [(Y - Y^d)' \vdots (X - X^d)']'$ and Q is negative definite and symmetric.[2] (3) is now rewritten as

$$H\tilde{Z} = b \tag{5}$$

where $H = [I \vdots -R]$ and $b = s - HZ^d$. Let $Q = \begin{bmatrix} B & C' \\ C & A \end{bmatrix}$ be conformable with $\tilde{Z}$.

We maximise (4) with respect to Z subject to the constraint set (5).

(4) and (5) are standard specifications in a linear-quadratic control problem. However they can be quadratic and linear approximations to more complicated nonlinear specifications — as for example in each step of Rustem, Vellupillai and Westcott (1977) where neither the weights nor the functional form of (4) can be specified to any acceptable accuracy; or in each successive linearisation in the algorithms of say, Holbrook (1974) or Preston *et. al.* (1976) for nonlinear models. This is a particularly important point since the thrust of this paper is to assess the changes in the optimal $\tilde{Z}^*$ resulting from perturbations in β, Γ, Q or q. Such perturbations will naturally result from the successive approximations, as those parameters will then depend on the levels of $\tilde{Z}$ in the current iterate of the solution.

III. RECONCILING VARIOUS ALGORITHMS

Algorithms for the four best known solutions to our control problem are reconciled into one notation so that the subsequent analysis may be applied equally easily to any one of them. Not only is this a simplification but it is in fact necessary as the analysis proved intractable using the original forms.

(a) A Lagrangian solution: Define Lagrange multipliers $\lambda = [\lambda_1' \ldots \lambda_T']'$ where $\lambda_i' = (\lambda_{1i} \ldots \lambda_{mi})$. We have to maximise

$$L = \tfrac{1}{2}\tilde{Z}'Q\tilde{Z} + q'\tilde{Z} - \lambda'(H\tilde{Z} - b) \tag{6}$$

Solving the first order conditions for the unknowns $\tilde{Z}$ and λ at the maximum gives

$$\begin{aligned} \lambda &= (HQ^{-1}H')^{-1}(b + HQ^{-1}q) \text{ and} \\ \tilde{Z}^*_\lambda &= Q^{-1}[H'(HQ^{-1}H')^{-1}(HQ^{-1}q + b) - q] \end{aligned} \tag{7}$$

(b) A substitution approach (Theil (1964)): substitute (5) into (4), writing $H^* = [R' \vdots I]$, and maximise the result. We have

$$\tilde{Z}^*_S = \begin{bmatrix} I_m & R \\ 0 & I_n \end{bmatrix} \left[\begin{bmatrix} I_m \\ -(H^*QH^{*\prime})^{-1}H^* \begin{bmatrix} B \\ \hline C \end{bmatrix} \end{bmatrix} b + \begin{bmatrix} 0 \\ -(H^*QH^{*\prime})^{-1}H^*q \end{bmatrix} \right] \tag{8}$$

Now

$$\begin{bmatrix} I & R \\ 0 & I \end{bmatrix}^{-1} = \begin{bmatrix} H \\ 0 \vdots I_n \end{bmatrix}; \; H^*H' = 0;$$

writing $\tilde{Z}^*_\lambda = \theta_i b + \theta_2 q$ and $\tilde{Z}^*_S = \theta^*_1 b + \theta^*_2 q$ it is easily verified that

$$\begin{bmatrix} H^*QH^{*'} & 0 \\ 0 & H^*QH^{*'} \end{bmatrix} \begin{bmatrix} H \\ 0 \vdots I \end{bmatrix} (\theta_1 - \theta_2)$$

$$= \begin{bmatrix} H^*QH^{*'} & 0 \\ 0 & H^*QH^{*'} \end{bmatrix} \begin{bmatrix} H \\ 0 \vdots I \end{bmatrix} (\theta_2 - \theta^*_2) = 0 \tag{9}$$

where both the premultiplying matrices are nonsingular.

Our control problem is stochastic, therefore we maximise the expectation of (4) and use the well known certainty equivalence result

$$\text{i.e.} \max_{\tilde{Z}} \{E(w) | H\tilde{Z} - b = 0\} = \max_{\tilde{Z}} \{w | H\tilde{Z} - E(b) = 0\} \tag{10}$$

(1) is dynamic, and $E(b)$ alters at each period t as new information in the form of actual residuals U_i and N_i, $i = 1, \ldots, t-1$ has revealed itself. So we replace $E(b)$ in (10) by $E_t(b)$ – the conditional expectation at t – and recalculate $\tilde{Z}^*_s$ to the remaining $(T-t)$ horizon. The T–period problem becomes T problems of periods $(T-t)$ in each of which only the first period $\tilde{Z}^*_s$ plan is implemented, and the remainder is the current best forecast of what the plans will be to the horizon. Since (7) and (8) are identical this first period certainty equivalence is appropriate also for (7).

For each of the $T-1$ solutions with horizon $(T-t)$, $t = 1, \ldots, T-1$ we have the additional constraint that $(\tilde{X}^{*'}_1 \ldots \tilde{X}^{*'}_{t-1})'$ is now fixed at its multiperiod certainty equivalent value. We add to b_t the term

$$\hat{R}_t \begin{bmatrix} \tilde{X}^*_1 \\ \vdots \\ \tilde{X}^*_{t-1} \end{bmatrix} \quad \text{where } \hat{R}_t = \begin{bmatrix} R_t & \cdots\cdots & R_2 \\ \vdots & & \vdots \\ R_T & \cdots\cdots & R_t \end{bmatrix}$$

in the calculation of the next $(\tilde{X}^{*'}_t \ldots \ldots \tilde{X}^{*'}_T)'$ (Theil (1964)). Overall the multiperiod centainty equivalent solution is the list of first period certainty equivalent solutions from the problems with horizons $T, \ldots, 1$.

In that list we have used $E_t(b_t) = E_t(s_t) - H_t Z_t^d + R_t^+ \tilde{X}^*$ for $t = 1, \ldots, T$ where $R^+ = R - (\text{diag } R_1)$ and H_t holds the rows and columns $mt+1, \ldots, mT$ from H. The subscript t on R^+, b, s, Z^d indicates that only rows $mt+1$ to mT are used. The arrays have elements deleted appropriately for each reduction in the horizon.

(c) Dynamic programming (Chow[3] (1975), Wall and Westcott (1975)). These algorithms have block diagonal A, B, C (and $q = 0$) so that no weights are placed on $\tilde{Z}_t' \tilde{Z}_{t'}$, $t \neq t'$. Only a sequence of optimal plans to be actually implemented is calculated. This sequence is identical to the first period certainty equivalent plans above, with the successive 'forecasts' discarded. Taking (8) to calculate the optimal instruments for $t = T$ we have,

$$-H_T^* Q_T H_T^{*\prime} \tilde{X}_T^* = H_T^* Q_T \begin{pmatrix} b_T \\ 0 \end{pmatrix} \tag{11}$$

where $Q_T = \begin{bmatrix} B_T & C_T' \\ C_T & A_T \end{bmatrix}$ in obvious notation. This is identical to the dynamic programming solution

$$X_t^* = L_t Y_{t-1}^* + h_t \tag{12}$$

(in either reference[4]) upon substitution for $h_t = L_t \Pi_1^{-1} (\Pi_3 N_T)$ and $\tilde{Y}_{T-1}^*$ from (2), and writing $H_T^* = (\Pi_{20} \vdots I)$. For $t = T-1$, (8) sets $\tilde{X}_{T-1}^*$ by the solution of simultaneous matrix equations as

$$\begin{aligned}
&-\{A_{T-1} + R_1' C_{T-1}' + C_{T-1} R_1 + R_1' B_{T-1} R_1 + R_2'[B_T - (C_T' \\
&+ B_T R_1)(R_1' B_T R_1 + A_T + C_T R_1 + R_1' C_T')^{-1}(R_1' B_T + C_T)] R_2\} X_{T-1}^* \\
&= [A_{T-1} + R_1' C_{T-1}' + C_{T-1} + R_1' B_{T-1}] b_{T-1} + R_2'[B_T - (C_T' \\
&+ B_T R_1)(R_1' B_T R_1 + A_T + C_T R_1 + R_1' C_T)^{-1}(R_1' B_T + C_T)] b_T
\end{aligned} \tag{13}$$

which again is identical to the dynamic programming solution for $T-1$ in (12) (substituting for b_T, b_{T-1}) since $R_2 = \Pi_1 R_1 + \Pi_{21}$ and by definition of $E_t(b_t)$. For every $t = 2, \ldots, T-2$ we repeat from (11) to (13), at each step, where we label R_1 and subscript $T-1$ to refer to items of the first period solution and R_2 and subscript t to those of the 'forecast' solutions to be discarded and the previous step respectively. Thus by induction (8) and (12) are identical if only the first period solutions of (8) are retained.

(d) The Pontryagin maximum principle solution: if there are no transversality conditions then a discrete version of the maximum principle using multiperiod certainty equivalence is identical to the dynamic programming solution, and therefore (8), because the Hamiltonian coincides with the sequential constrained optimisations, as pointed out by Chow (1975).

Therefore each decision rule above for $\tilde{Z}^*$ is identical for any given welfare function and model subject to the stated restrictions. Moreover each separate decision rule may be written identically, so that for each any analysis can be conducted equally well on one of them alone.

IV. MULTIPERIOD CERTAINTY EQUIVALENCE

The reconciliation of these decision rules was obvious because (4) has a unique constrained maximum. We may now conclude:

(a) $\tilde{Z}_t^*$ from (12) is a multiperiod certainty equivalent solution, and therefore so is the sequence of first period solutions from (7) or (8), when A, B and C are block diagonal. This follows since

$$L_t = -[\Pi_{20}' W_t \Pi_{20}]^{-1} \Pi_{20}' W_t \Pi_1 \text{ and}$$
$$h_t = -[\Pi_{20}' W_t \Pi_{20}]^{-1} \Pi_{20} W_t (\Pi_3 N_t - g_t)$$

where $W_{t'}$ W_{t-1} and $g_{t'}$ g_{t-1} are contained in (11) and (13) respectively. Therefore $\tilde{X}_t^* = (I - P_t)(\tilde{Y}_t + g_t)$ and

$$\tilde{Y}_t^* = P_t \tilde{Y}_t + (I - P_t) g_t \tag{14}$$

where $P_t = I - (\Pi_{20}' W_t \Pi_{20})^{-1} \Pi_{20}' W_t$. Thus, comparing $E(w(\tilde{Z}))$ $= \sum_t E_t(w(\tilde{Z}))$ and $w(E(\tilde{Z})) = \sum_t w(E_T(\tilde{Z}_t))$ for some $\tilde{Z}_t$ the difference is $\frac{1}{2}\sum_t [E_t b_t' Q_t b_t) - (E_t b_t)' Q_t (E b_t)]$ (using $\tilde{Y}_t = \sum_{i=1}^{t} R_i \tilde{X}_i + b_t$) which is independent of $\tilde{Z}$. Moreover, it is separable over time so in particular the maximum is computed from $\max_{\tilde{Z}_t} \; w(E_t(\tilde{Z}_t))$, $t = 1 \ldots T$.

(b) A certainty equivalent solution for $\tilde{X}_t^*$ at its time of execution, depends on $E_t(b_t)$ which is unavailable before t. For a general Q, an alternative solution of the style of (12) is required in which $\tilde{X}^*$ is successively reevaluated for all remaining time periods simultaneously. The computation appears very complicated. Simpler is (8) (with 'forecasts' discarded) which is multiperiod certainty equivalent for a general Q since

$$\sum_{t=1}^{T} E_t(w(\tilde{Z}^{(t)})) - \sum_{t=1}^{T} w(E_t \tilde{Z}^{(t)}) = \sum_{t=1}^{T} \sum_{i=t}^{T} \sum_{j=t}^{T} \{ E_t(b_1' Q_{ij} b_j) - (E_t b_i') Q_{ij} (E_t b_j) \} \tag{15}$$

which is independent of $\tilde{Z}$ and separable over time, for some $\tilde{Z}^{(t)}$. $\tilde{Z}^{(t)}$ is some extended vector of length $(m+n) \sum_{i=1}^{T-t+1} i$ and Q_{ij} a submatrix of Q. These sequential plans are essential to multiperiod certainty equivalence.

V. SENSITIVITY TO STOCHASTIC SHOCKS

The decision rules (7) and (8) are simple to work with since so far the stochastic elements are confined to b. (8), by avoiding Q^{-1} involves less computation. We have[5] for horizon T

$$\frac{\partial Z^*}{\partial U_{jt}} = \begin{bmatrix} I & R \\ 0 & I \end{bmatrix} \begin{bmatrix} I \\ -(H^*QH^{*\prime})^{-1}H^* \begin{bmatrix} B \\ C \end{bmatrix} \end{bmatrix} S \frac{\partial U}{\partial U_{jy}} = \theta_i^* S \frac{\partial U}{\partial U_{jt}} \tag{16}$$

$$t = 1, \ldots, T; j = 1 \ldots m^*.$$

The sequence of T solutions, with horizons $t = T \ldots 1$, make use of a sequence of $\theta_1^{*(t)}$, in which R and Q are reduced in dimension by one row and column block of submatrices for each t.

Let $\theta_1^{*(t)} = \begin{bmatrix} \theta_Y^{*(t)} \\ \theta_X^{*(t)} \end{bmatrix}$ be conformable with that part of Z^* which remains to be implemented at t. Take $\theta_{11}^{*(t)}$ to be the first m rows of $\theta_Y^{*(t)}$ and θ_{12}^* to be the first n rows of $\theta_X^{*(t)}$ each followed by a null matrix dimension and $m \times m(T-t)$, $n \times m(T-t)$ respectively. Let also

$$R_{(t)}^{+} = \left[\begin{array}{cccc|c} 0 & & & & \\ & \ddots & & 0 & \\ R_2 & & \ddots & & 0 \\ \vdots & \ddots & & 0 & \\ R_{T-t} & \cdots & & R_2 & \\ \hline & \hat{R}_T & & & 0 \end{array}\right]$$

$\Omega_1{}^{(t)} = [\theta^*_{11}{}^{(1)'} \ldots \theta^*_{11}{}^{(t-1)'} \vdots\ \theta^{*(t)'}_Y]'$ and $\Omega_2{}^{(t)} = [\theta^*_{12}{}^{(1)'} \ldots \theta^*_{12}{}^{(t-1)'}$ $\vdots\ \theta^{*(t)'}_x]'$. Then for $t < T$

$$\frac{\partial Z^*}{\partial U} = \begin{bmatrix} \Omega_1{}^{(t)} \\ \Omega_2{}^{(t)} \end{bmatrix} \left[S + (0 \vdots R^+_{(t)}) \frac{\partial Z^*}{\partial U} \right]$$

so we have

$$\frac{\partial Z^*}{\partial U} = \left[\begin{matrix} I_{mT} \\ 0 \end{matrix} \vdots \begin{bmatrix} 0 \\ I_{nT} \end{bmatrix} - \begin{bmatrix} \Omega_1{}^{(t)} \\ \Omega_2{}^{(t)} \end{bmatrix} R^+_{(t)} \right]^{-1} \begin{bmatrix} \Omega_1{}^{(t)} \\ \Omega_2{}^{(t)} \end{bmatrix} S = G^{(t)}\Omega^{(t)}S \quad (17)$$

say (of order $(m+n)T \times m^* T$).

By similar steps we get sensitivity to exogenous information

$$\frac{\partial Z^*}{\partial \mathrm{N}_{jt}} = \theta^{*(t)}_1 \sum_{i=1}^{\tau} \Pi_1^{i-1} \Pi_3 \frac{\partial N}{\partial N_{j\tau}} \quad \tau = 1 \ldots T, j = 1 \ldots r, t = T - \tau \quad (18)$$

$$\frac{\partial Z^*}{\mathrm{vec}\ (\partial N)} = G^{(t)}\Omega^{(t)}R^* \qquad \text{(of order } (m+n)T \times rT) \qquad (19)$$

with

$$R^* = \begin{bmatrix} R^*_1 & & \\ & & 0 \\ R^*_T & & R^*_1 \end{bmatrix}$$

and

$$R^*_i = \Pi_1^{i-1} \begin{pmatrix} \Pi_3 \\ 0 \end{pmatrix}$$

The sequence of matrices $G^{(t)}$ $t = T, \ldots, 1$ is found[6] to be $G^{(t)} = I$ and leads to results for the multiperiod certainty equivalent Z^* in $\Omega^{(1)}S$ and $\Omega^{(1)}R^*$. Both (17) and (19) are matrices whose components will be computed automatically during the computation of $\hat{Z}^*$ itself. Some examples for the Theil problem are given in Tables 7A.1 and 7A.2.

(19) can be interpreted as recognising N may be stochastic—or at least uncertain. Bounds on likely future values indicate the possible variations in Z^*. Similarly (17) may be used with error variances to indicate possible variations due to the uncertainty in each equation of the model.

Finally we also have

$$\frac{\partial Z^*}{\partial Z^d} = (I - \Omega^{(t)} H) \text{ (square of order } (m+m)T) \tag{20}$$

The change in overall welfare resulting from any of these perturbations is important. Writing $Z^* = [RX^* + s \vdots X^*]$ and substituting into (4) we find

$$w^* = k - \tfrac{1}{2}\tilde{X}^{*\prime}(H^*QH^{*\prime})'\tilde{X}^* \text{ where } k = q'\begin{bmatrix} b \\ 0 \end{bmatrix} + \tfrac{1}{2}b'Bb \tag{21}$$

For any other X,

$$w = k + [\tfrac{1}{2}\tilde{X}' - \tilde{X}^{*\prime}]H^*QH^{*\prime}\tilde{X}^* \tag{22}$$

so with some rearrangement

$$w^* - w = -\tfrac{1}{2}(X - X^*)'H^*QH^{*\prime}(X - X^*) \tag{23}$$

Substituting $dX^* = X - X^*$ from the lower submatrices of (17), (19) or (20), as appropriate, in (23) indicates the welfare change from the current optimum which results from isolated alterations in either U, N or Z^d. But now results for the multiperiod certainty equivalent Z^* itself are obtained by direct substitution into (23); and changes attributable to individual periods by setting the remainder of $X - X^*$ to zero. We could also compare all or part of any of the sequential solutions with the final Z^* in (23) as a measure of the inferiority of the 'forecasts', or the benefit of the policy revisions.

VI. SENSITIVITY TO CHANGES IN THE MODEL PARAMETERS

The previous section may be regarded as treating sensitivities to information errors. In this section a method is proposed for calculating changes in Z^* due to perturbations in the model parameters—sensitivities to model errors. From (8) we could calculate, say $\dfrac{\partial \tilde{X}^*}{\partial R_{ij}}$ directly but the result is neither simple nor enlightening. And in any case this is not the problem which would interest the planner. He would need to assess the sensitivity of Z^* to small changes in one, several or possibly all elements of β, Γ. Even a single parameter change will imply every element of R changes. These changes might be the result of reestimation of the model, new data becoming available, alternative estimates or specifications of the model being used; but most likely the planner will chiefly be interested in the effects on Z^* of the possible variations in the estimated elements of β, Γ due to their uncertainty and stochastic nature. He might put bounds on the changes to Z^* when each parameter varies between the bounds of twice its standard error individually and collectively.

A direct calculation of these sensitivities, although possible by repeated Monte Carlo experiments, would be very complicated, and probably uninformative. Alternatively

$$(\partial\Pi_1 \vdots \partial\Pi_2 \vdots \partial\Pi_3) = \beta^{-1}(\partial\beta \vdots \partial\Gamma)\begin{bmatrix}\Pi_1\Pi_2\Pi_3 \\ I\end{bmatrix}; \frac{\partial R_i}{\partial\Pi_1} \text{ from } (i-1)R_{i-1}$$

$$\frac{\partial R_1}{\partial\Pi_1} = 0; \frac{\partial R_i}{\partial\Pi_2} \text{ from } \Pi^{i-1}; \text{ and } \frac{\partial\tilde{X}^*}{\partial R} \text{ from (8) etc.} \tag{24}$$

shows something of the complications of these calculations. Instead we can use, from (17) and (1) (for $t = 1 \ldots T$)

$$\frac{\partial Z^*}{\text{vec}\,\partial(\beta \vdots \Gamma)} = \frac{\partial Z^*}{\partial U}\frac{\partial U}{\text{vec}\,\partial(\beta \vdots \Gamma)} = \Omega^{(t)}SD \text{ of order } (m+n)T \times (n^*+r+m^*)^2 \tag{25}$$

where $D = [\text{diag } D^{(i)}]$ and the submatrix $D^{(i)}$ has typical element

$$\left.\begin{aligned} D_{tj}^{(i)} &= Y_{jt}^* \text{ if } \beta_{ij} \neq 0 \\ D_{t(k+m^*)}^{(i)} &= \overline{X}_{kt}^* \text{ if } \gamma_{ik} \neq 0 \end{aligned}\right\}$$

for

$$i = 1, \ldots, m^*; j = 1, \ldots, m^*$$
$$k = 1, \ldots, n^*+m^*+r \text{ and } t = 1, \ldots, T$$
and zero elsewhere

In other words we find changes in the structural errors which are equivalent to the given changes in $(\beta \vdots \Gamma)$, (keeping the variables constant) and use those equivalent changes in (17). We could have done the same using (19) and $\partial N/\partial(\beta \vdots \Gamma)$ in (25) but usually there are fewer elements in N_t than in U_t; and these parameter changes are almost certain to be confined to the stochastic equations rather than the identities. However we could model changes in Z^* using both U and N if necessary. The disadvantage of this method is we do not calculate the actual changes in R. But the combined effect on a single element of Z^* of several parameter changes can be obtained from (25).

(25) is very simple to use. Some selected results from the Theil example is given in Tables 7A.3a and 7A.3b. The welfare change consequent on individual parameter changes can be calculated from (23).

VII. CHANGES IN THE WELFARE FUNCTION PARAMETERS

Apart from an interest in changes in Z^* due to successive reevaluations of a quadratic approximation to a nonquadratic preference function, the planner will need to assess the sensitivity of Z^* to changes in some or all of the elements of Q and q when he is uncertain of the precise specification of his relative preferences over targets and/or instruments and their values over time. It will help him decide where to concentrate his attention in that specification or, in the case of the respecification algorithm, on which of the deviations from desired values of the variables he should first concentrate. The method proposed below can also be used to assess the effects of introducing a new target or instrument to the problem. It is important to consider the latter because if the planner is uncertain about his relative weights then he will also be uncertain about the list of targets and instruments. Casual observation clearly reveals this to be the case in the practical world of governments and planners.

It is convenient to use (7) for this section. We obtain the element i, j;

$$\frac{\partial Z^*}{\partial Q_{ij}} = -[I - Q^{-1}H'(HQ^{-1}H')^{-1}H]Q^{-1}\frac{\partial Q}{\partial Q_{ij}}\tilde{Z}^* \qquad (26)$$

and

$$\frac{\partial Z^*}{\partial q_k} = [I - Q^{-1}H'(HQ^{-1}H')^{-1}H]Q^{-1}\frac{\partial q}{\partial q_k} \qquad i, j, k = 1 \ldots (n+m)T \qquad (27)$$

So

$$\frac{\partial Z^*}{\partial Q} = \tilde{Z}^{*\prime} \otimes \phi \text{ and } \frac{\partial Z^*}{\partial q} = \phi$$
$$\text{where } -[I - Q^{-1}H'(HQ^{-1}H')^{-1}H]Q^{-1} = \phi \qquad (28)$$

Since Q is symmetric and negative definite $Q^{-1} = -K'K$ where K is triangular and nonsingular. Hence

$$\phi = K'[I - K^{*\prime}(K^*K^{*\prime})^{-1}K^*]K = K'MK \text{ where } K^{*\prime} = KH' \qquad (29)$$

Therefore $\frac{\partial Z_j^*}{\partial Q_{ij}}$ has the same sign as Z_j^* all $i = 1, \ldots, (n+m)T$ for each $j = i, \ldots, (n+m)T$ (or to Z_j^* if $q = 0$) and similarly $\frac{\partial Z_k^*}{\partial q_k} \leqslant 0$ $k = 1, \ldots, (n+m)T$(M is indempotent).[7] The easiest way to compute $-\phi$ is as the variance-covariance matrix of residuals in the regression[8] of K on H'. (28) may also be obtained from $\phi = -(I - \Theta_i H)Q^{-1} = (\Theta_1^* H - I)Q^{-1}$.

Selected values for these sensitivities are computed in Table 7A.4 – again for the Theil example. Results for the multiperiod certainty equivalent solutions will be in the first $(n+m)$ rows and columns in the four blocks of (28) corresponding to the partitioning of Q in each successive computation of those matrices, each with T reduced by one. As before the welfare changes are available from (23).

VIII. TOTAL CHANGES IN Z^*

The total change in any element of Z^* is given by

$$dZ_i^* = \sum_k \frac{\partial Z_i^*}{\partial U_k} dU_k + \sum_j \sum_k \frac{\partial Z_i^*}{\partial N_{jk}} dN_{jk} + \sum_j \sum_k \frac{\partial Z_i^*}{\partial (\beta \vdots \Gamma)_{jk}} d(\beta \vdots \Gamma)_{jk}$$
$$+ \sum_k \frac{\partial Z_i^*}{\partial Z_k^d} dZ_k^d + \sum_j \sum_k \frac{\partial Z_i^*}{\partial Q_{jk}} dQ_{jk} + \sum_k \frac{\partial Z_i^*}{\partial q_k} dq_k \quad i = 1, \ldots, (n+m)T \tag{30}$$

when all parameters in the economic specification of our problems are allowed to vary. The multiperiod certainty equivalent sensitivities are obtained from $i = 1 \ldots (n+m)$ in (30) from each member sequence $(T \ldots 1)$ of such expressions. (30) is easily computed from (17), (19), (25), (20) and (28). Obviously the effects of simultaneous changes in a subset of the parameters—for example variations in several or all of the model's structural form coefficients—are obtained from (30). For the complete control problem we have

$$dZ^* = dZ^d + \Omega^{(1)}[S(dU + D \text{ vec } d(\beta \vdots \Gamma)) + R^* \text{ vec } dN - HdZ^d] - \phi dq$$
$$+ Z^{*\prime} \otimes \phi \text{ vec } dQ \tag{31}$$

and welfare changes from (23). One interesting consequence of (31) is that the multiperiod certainty equivalent solution can now very simply be computed in one operation (in the same sense as (12)) without computing, and then discarding, the 'forecast' values for a given general (Q, q). If $\tilde{Z}^{*(1)}$ is the first period certainty equivalent solution for horizon T then

$$\tilde{Z}^* = \tilde{Z}^{*(1)} + \Omega^{(1)} S^+ dU \text{ where } S^+ = S - (\text{diag } S_1) \qquad (32)$$

The difference between corresponding elements in two of the $T(m+n)$–subvectors of the last two terms of (32) show the optimal revision to plans of the later period over those of the earlier. (32) assumes the uncertainty is only in U but if, as is highly likely, values taken for N_t are revised when the actuals become known then we add $\Omega^{(1)} R^{*+}$ (vec dN) on the right. Here the revisions (deviations of actuals from expected values) in U and N are dU and vec dN respectively and $R^{*+} = R^* -$ (diag R_1^*).

IX. CHANGING THE LIST OF TARGETS AND INSTRUMENTS

We now apply (28) to assess the effects on Z^* of the removal or inclusion of an instrument or target from the control problem. For removal, we use (28) with $dQ_{ij} = -Q_{ij}$, where $i = kt, j = 1 \ldots (n+m)T$ and $Q_{ij} = Q_{ji}$ for $t = 1 \ldots T$ and k being the variable in Z_t being omitted, and compute

$$dZ^* = \frac{\partial Z^*}{\partial Q} \text{vec } (dQ) \qquad (33)$$

For inclusion, we assume (8), say, was computed with $Q_{ij} = 0$, $i = kt$, $j = 1 \ldots (n+m)T$, $t = 1 \ldots T$ and k being the variable to be included. Then (28) and (33) are used where dQ_{ij} is the weight we now wish to attach to the included variable for each $t = 1 \ldots T$. The benefits or costs to these changes in the list of targets and instruments may be assessed from (23) and dZ^*.

Some particular cases will be of special interest. For example, the omission or inclusion of subsets of targets and/or instruments is likely to have quite different consequences from the omission or inclusion of members of those subsets one at a time. This proposition can be tested in a given case, and the extent of these differences measured by the alterations to elements of Z^*. Apart from the overall effects as measured by (23), we may be interested in the differences in various values of particular elements of dZ^* at various points of time. Thus some measures of the importance of the careful coordination of two or more policies, both in their overall effects and of their specific incidence on other variables, are available. Similarly by setting the above Q_{ij} nonzero and zero for only some of $t = 1 \ldots T$ we get similar measures of the importance of timing for a particular policy. However these measures of importance are crude. A proper analysis must investigate the substitutability or complementarity of particular policies, and their own values over time.

Finally new instruments, not already in the model, may be introduced by reestimation of the appropriate equation and then using (25). D is set up to accept new variables by adding column, or for changing the specification of existing equations by inserting newly estimated terms in the appropriate zero position.

X. RESULTS

The results given show $\pm dZ^*$ values resulting from changes in various variables. Signs are therefore ignored, except in Tables 7A.1 and 7A.4 for the individual partial derivatives.

In summary, the dominant conclusion is that Z^* is markedly less sensitive to changes in elements of Q than to equivalent sized shocks to the model parameters or equation errors. Confidence intervals for these shocks show that Z^* is likely to be more disturbed by them than by equivalent changes in Q elements. Comparing Tables 7A.1 and 7A.3 with 7A.4, individual unit changes in Q and U, and much smaller ones in $(\beta; \Gamma)$, produce dZ^* values of similar absolute size; but likely values of dU or $(\beta;$ $\Gamma)$ produce dz^* larger in absolute size than for dQ of the same order. Table 7A.4 shows only the largest values which, when compared with the largest values for the same variables in Tables 7A.1 and 7A.3, are smaller. And these large values are very few in Table 7A.4 (note the mean and standard error figures) and noticeably less numerous than those of Tables 7A.1 and 7A.3. I, D and G appear most often affected in Table 7A.4; D significantly. In Tables 7A.1 and 7A.3 those variables and W_2 are most affected but D less than in Table 7A.4.

That conclusion suggests more effort be put into accurate model specification and estimation, and less into the fine detail and precision of the relative weights in (4). Probably only a few points need careful specification. Within reason, therefore, roughly specified preferences and trade-offs give similar results to those from a very careful specification—both in optimised variables and welfare index. This is a useful conclusion—in so far as it extends to other problems—because the specification of Q is the area of our least experience and knowledge. It is often argued that there is no hope of accurate specification—even from the revealed preferences of the planning authority—so that we should

TABLE 7A.1 Sensitivities as a % of optimised value (bracketed figures absolute change, $b constant prices)
A. Sensitivity to unit individual stochastic shocks, first periods solution, $T = 4$.

	C	I	D	w_2	T	G
U_C	1.63 (.8098)	.57 (.02717)	.47 (.03147)	−8.24 (−.5053)	2.89 (.2488)	−4.97 (−7122)
U_I	.15 (.0748)	21.6 (1.0383)	.67 (0.3207)	−1.35 (−.08328)	3.88 (.3343)	−3.41 (−.4867)
U_w	1.53 (.75698)	2.14 (.10254)	9.2 (.6237)	−9.1 (−.5574)	−8.84 (−.7617)	−1.7 (−.2429)
	B. Sensitivity to individual stochastic shocks at ± 2 SE (5 % probability bounds)					
U_C	3.68 (1.85)	1.28 (.062)	1.08 (.07)	18.79 (1.149)	6.58 (.568)	11.33 (1.62)
U_I	.36 (.178)	51.3 (2.46)	1.13 (.076)	3.2 (.197)	9.2 (.79)	8.14 (1.16)
U_w	1.89 (.925)	2.6 (.124)	11.22 (.76)	14.13 (.87)	13.75 (1.186)	2.64 (.38)
(*a*)	6.47 (3.21)	56.0 (2.69)	18.05 (1.22)	36.2 (2.22)	29.4 (2.54)	22.15 (3.16)
(*b*)	2.9 (1.44)	25.2 (1.21)	8.12 (.549)	16.15 (.99)	13.2 (1.14)	9.96 (1.42)
(*c*)	2.2 (1.079)	27.3 (1.31)	2.8 (.19)	12.4 (.76)	9.9 (.86)	9.81 (1.4)
(*d*)	1.19 (.57)	25.6 (1.23)	1.2 (.079)	5.2 (.32)	.42 (0.36)	.66 (0.94)

(*a*) The maximum change if each error is at ± twice its SE (no cancelling) – probability at most 5.2×10^{-6}.
(*b*) The maximum change with each error at ±.9 its SE (no cancelling) – probability at most 1.5 %.
(*c*) As (*b*) but orthoganalised errors to allow for covariances of U_C, U_I, U_w – with probability at most 5 %.
(*d*) The maximum change with orthoganalised errors at ±.9 SE (maximum cancelling and probability at most 5 %).

TABLE 7A.2 Sensitivity to errors in uncontrollable variables. (10 % changes in N for first period solution, $T = 4$.)

	C	I	D	w_2	T	G
t	.033 (.02)	.91 (.045)	.175 (.013)	.24 (.015)	.23 (.02)	.045 (.007)
constant	3.9 (1.93)	39.8 (1.91)	3.1 (.21)	16.3 (1.08)	13.1 (1.13)	14.6 (2.09)

TABLE 7A.3a First period solution, $T = 4$. Sensitivity to errors in structural parameters: 10% individual parameter changes.

	C	I	D	w_2	T	G
α_0	2.69 (1.36)	1.13 (.055)	.79 (.054)	13.5 (.83)	4.8 (.41)	8.36 (1.19)
α_1	.035 (.018)	.2 (.008)	.009 (.0007)	.14 (.011)	.05 (.006)	.11 (.016)
α_2	.27 (.13)	.09 (.005)	.07 (.005)	1.08 (.08)	.44 (.04)	.76 (.112)
α_3	4.29 (2.15)	1.86 (.09)	1.31 (.09)	22.4 (1.37)	7.9 (6.8)	13.5 (1.94)
β_0	2.67 (.133)	38.4 (1.85)	.84 (.06)	2.4 (.14)	5.03 (.59)	6.06 (.86)
β_1	.039 (.019)	5.52 (.27)	.12 (.008)	.35 (.02)	1.0 (.087)	.87 (.125)
β_2	.054 (.028)	8.0 (.39)	.18 (.012)	.52 (.03)	1.5 (.13)	1.3 (.186)
β_3	.44 (.225)	63 (2.98)	1.39 (.09)	3.96 (.24)	11.3 (.98)	10.0 (1.43)
γ_0	.33 (0.16)	.30 (.016)	1.46 (.099)	1.44 (.09)	1.43 (.122)	.27 (.039)
γ_1	2.86 (1.43)	3.95 (.19)	7.3 (.49)	17.2 (1.06)	16.7 (1.44)	3.21 (.46)
γ_2	1.16 (.55)	1.45 (.07)	6.5 (.44)	6.6 (.14)	6.36 (.55)	1.75 (.25)
γ_3	.03 (.02)	.06 (.007)	.26 (.02)	.2 (.02)	.23 (.021)	.064 (.009)

for other % changes multiply the above appropriately.

TABLE 7A.3a (*continued*) Individual parameters shifted to ± twice their SE (at most 5% probability)

	C	I	D	w_2	T	G
α_0	3.93 (1.98)	1.38 (.066)	1.19 (.08)	21.3 (1.26)	7.3 (.63)	12.7 (1.82)
α_1	3.49 (1.79)	1.21 (.06)	1.02 (.069)	18.4 (1.12)	6.4 (.55)	11.04 (1.53)
α_2	4.72 (2.4)	1.62 (.078)	1.49 (.1)	25.1 (1.5)	8.8 (.76)	15.1 (2.1)
α_3	1.64 (.83)	.58 (.03)	.5 (.03)	4.4 (.27)	3.1 (.26)	5.3 (.76)
β_0	2.17 (1.08)	324 (14.9)	6.8 (.46)	1.89 (.12)	5.5 (.147)	4.85 (.69)
β_1	.53 (.26)	73.9 (3.55)	1.63 (.11)	4.7 (.29)	8.4 (.71)	11.97 (1.71)
β_2	.30 (.15)	42.7 (2.05)	.95 (.06)	2.75 (.17)	7.9 (.68)	6.98 (.995)
β_3	2.09 (1.04)	374 (14.6)	6.67 (.45)	19.0 (1.17)	54.4 (4.7)	47.9 (6.85)
γ_0	.41 (.23)	4.83 (.23)	208 (1.41)	2.04 (.13)	2.12 (.18)	.39 (.06)
γ_1	5.06 (2.5)	6.9 (.33)	12.9 (.87)	30.3 (1.86)	29.0 (2.52)	5.7 (.81)
γ_2	5.2 (3.05)	6.9 (.33)	30.6 (2.08)	31.2 (1.92)	30.6 (2.64)	5.8 (.82)
γ_3	.16 (.06)	0.65 (.03)	1.02 (.07)	.96 (.06)	.97 (.086)	.26 (.03)

TABLE 7A.3b

		C	I	D	w_2	T	G
α	(a)	5.03(2.51)	1.75(.84)	1.5(.101)	15.9(.976)	9.28(.799)	16.05(2.3)
	(b)	1.53(.77)	.53(.03)	.397(.027)	1.45(.08)	1.02(.088)	1.77(.25)
	(c)	3.48(1.75)	1.3(.06)	1.02(.07)	17.7(1.09)	6.24(.54)	10.8(1.54)
	(d)	.115(.058)	.14(.002)	.03(.002)	.58(.04)	.21(.018)	.36(.052)
β	(a)	1.9(.95)	275(13.16)	5.99(.406)	10.65(.65)	28.6(2.47)	26.9(3.85)
	(b)	.14(.07)	23.4(1.14)	.53(.076)	3.63(.22)	12.24(1.05)	9.09(1.29)
	(c)	1.09(.55)	158(7.58)	3.46(.234)	9.92(.16)	28.3(2.44)	24.9(3.55)
	(d)	.47(.23)	66.8(3.20)	1.46(.099)	4.19(.25)	11.97(1.03)	10.5(1.5)
γ	(a)	4.01(2.02)	7.86(.348)	24.2(1.64)	23.9(1.47)	23.8(2.06)	4.52(.64)
	(b)	.16(.089)	4.22(.201)	1.89(.125)	.78(.047)	.68(.06)	.14(.023)
	(c)	2.7(1.34)	3.77(.18)	16.2(1.1)	16.05(.98)	15.6(1.35)	3.0(.43)
	(d)	.34(.173)	.48(.02)	2.1(.145)	2.1(.13)	2.0(.17)	.39(0.56)

(a) The maximum dZ^* when each parameter of one equation is at $\pm$ .718 of its SEs (no cancelling) – probability at most 5%.
(b) as (a) but maximum cancelling.
(c) as (a) using orthogonalised parameters – probability at most 5%.
(d) as (b) using orthogonalised parameters.

All parameters varying simultaneously: orthogonalised parameters (no cancelling) – probability at most 5%.

C	I	D	w_2	T	G
2.87(1.4235)	63.8(3.0625)	8.15(.551)	17.1(1.052)	19.6(1.694)	15.2(2.162)

TABLE 7A.4

A. Sensitivity to unit changes in elements of Q, solution for $T = 4$ first period solution, sensitivities exceeding $|.6|$

variable:	*C*	*I*	*D*	*D*	*D*	
Q row	1	1	2	2	3	there are no other multiperiod certainty equivalent solution members in this class
Column	3	3	2	3	3	
Change as % of optimised variable	−2.2	−29.1	−14.2	−41.8	22.8	
Absolute change	−1.1	−1.4	−.96	−2.82	1.54	

B. The other first period solutions:

Variable:	*I*	*I*	*D*	*D*	*D*	*D*	*D*	*D*	*G*	*G*	*G*	*G*	*G*	*G*	*G*
Q row	1	1	2	3	2	3	2	2	8	8	8	8	8	8	8
Column	7	10	7	7	10	10	4	9	1	2	3	6	9	11	12
% change	−30.2	−31.2	21.7	10.6	23.6	11.1	15.9	−15.8	4.7	5.3	16.8	9.6	10.5	−5.9	−6.7
Absolute change	−1.45	−1.5	1.47	.72	1.6	.75	1.08	−1.07	.68	.75	2.4	1.38	1.5	−.84	−.96

TABLE 7A.4 *(continued)*

C. The largest for the 'forecasts' $t = 2 \ldots . 4$, $T = 4$.

Variable (t)	$C(2)$	$C(4)$	$w_2(2)$	$w_2(3)$	$w_2(4)$	$T(3)$	$T(4)$	$G(3)$
Q row	1	5	9	10	12	10	12	12
Column	2	10	12	10	12	10	12	12
% change	3.1	3.6	−95.0	24.5	−30.5	13.9	−46.5	15.6
Absolute change	1.59	2.3	−5.95	1.51	−1.44	1.06	−3.25	2.14

D. Total elements of ϕ exceeding $|.6|$ is 8 out of 576 i.e. 1.4%, less than $|.1|$ is 465 out of 576 i.e. 80.7%.

	C	I	D	w_2	T	G	at $t = 4$
$\|.6\|$ represents	.1 (.07)	20 (.96)	39 (2.7)	10.8 (.66)	8.8 (.76)	15.8 (2.2)	
but $\|.1\|$ represents	.02 (.012)	3.3 (.16)	6.6 (.45)	1.8 (.109)	1.4 (.13)	2.6 (.37)	

E. Mean and standard error sensitivities for the first period solution $T = 4$.

C −.01 (−.007) and .02 (.01)	D −2.68 (−.0961) and 39.6 (1.4)	T +.9 (+.0181) and 10.4 (.21)
I −1.31 (−.0437) and 27.1 (.9)	w_2 +.7 (+.0415) and 11.4 (.68)	G +2.75 (+.081) and 19.2 (.57)

abandon the exercise or simulate with various welfare functions.

Secondly Z^* is more sensitive to parameter shocks than equation error shocks at equivalent probability levels. This conclusion is expected since (25) indicates dZ^* is the product of the sensitivity, the parameter disturbance and the current value of the associated variable. Thus it will only be true when the likely variation in the parameter is less than that of the equation error by more than that variable's value. Z_i^* is therefore sensitive to parameter shocks in equations to whose errors it is sensitive, and where the associated variable takes relatively large values. The evidence is in comparing Tables 7A.1 with 7A.3 and 7A.5a with 7A.5b – and, for the reasons given, will inevitably extend to other economic control problems.

The details of Table 1 give information about the behaviour of Z^* under stochastic shocks, with implications for the contribution of individual equations. Broadly, positive dU increases Y^* and depresses X^*, but vice versa, for positive dQ (in the current period). Table 7A.6 confirms this for subsequent periods. Sensitivities are rather similar sizes for each error, but targets are more sensitive than instruments in absolute values. Characterising likely shocks by standard errors, the errors produce similar absolute disturbances—perhaps U_I most and U_C least for the targets and vice versa for instruments; C, I and G are most affected. In relative disturbances, U_W affects the targets most (apart from U_I on I); and U_C and U_W the instruments most. The disparity of results for relative and absolute disturbances is partly due to weighting all absolute deviations equally. On this basis it is rather unclear where disturbances to Z^* are worst and therefore respecification is most urgently required (beyond the implications of results mentioned above). However Z^* as a whole is fairly robust.

The reliability of Z^* cannot be judged from considering single random shocks. Tables 7A.1 and 7A.5a show dZ^* for all errors varying simultaneously. They show I, of the targets, least robust both absolutely and relatively, but of the instruments G absolutely and W_2 relatively. Table 7A.1 has been set up to show the effects of individual and combinations of errors as components of dZ^* for some given probability levels. We have from it a number of values of dZ_i^* occuring with equal probability being on an ellipsoid surface whose axes are the values of the individual errors. Due to possible cancelling of random error effects in dZ^* the ellipsoid represents the 'worst' situation for each component and their combinations in the error space. For the same probability level there is a 'best' situation—giving a star shaped surface contained within the ellipsoid and tangent to it where the ellipsoid gradient is zero or

TABLE 7A.5a Z_i^* as a function of orthogonalised random equation errors; probabilities of dZ_i^*.

Z_i^*	SE	$dZ_i^* = 0$	$< \pm 1$	$< \pm 2$	$< \pm .1Z_i^*$	$< \pm .5Z_i$	(e)
C	3.383	.1179	.233	.445	.8572	.9999	±3.96(1.97)
I	1.676	.2381	.440	.767	.2256	.8480	±59.66(2.86)
D	4.719	.0845	.170	.330	.1134	.5260	±5.4(.36)
w_2	1.932	.2065	.395	.699	.251	.887	±20.1(1.24)
T	4.027	.0991	.195	.383	.1704	.7154	±13.5(1.16)
G	5.923	.0674	.135	.264	.1916	.7716	±14.9(2.13)

TABLE 7A.5b Z_i^* as a function of orthogonalised random parameter errors; probabilities of dZ_i^* – first period solution $T = 4$.

Z_i^*	SE	$< \pm 1$	$< \pm 2$	$< \pm .1Z_i^*$	$< \pm .5Z_i^*$	(e)
C	1.9667	.39	.6922	.9882	.9999	±7.93(3.93)
I	7.8062	.1034	.2052	.0478	.2434	±325.0(15.6)
D	1.0634	.6528	.9398	.4778	.9986	±31.46(2.127)
w_2	1.4192	.519	.8414	.334	.9692	±17.4(2.83)
T	2.8453	.274	.5184	.2382	.871	±66.0(5.69)
G	3.882	.2034	.394	.2886	.9342	±54.4(7.76)

(e) is the 95% confidence interval as a % of the optimised variable together with the absolute value.

TABLE 7A.6 Sensitivities of the multiperiod certainty equivalent solution to unit stochastic shocks Horizon t = 4 results in Table 7A.1.

		w_2	T	G
Horizon t = 3	U_C	8.2(−.5055)	2.89(+.249)	4.99(−.7116)
	U_I	1.4(−.0837)	3.86(+.333)	3.41(−.486)
	U_w	9.1(−.558)	8.85(−.7613)	1.69(−.2415)
Horizon t = 2	U_C	8.3(−.5089)	2.93(+.2522)	4.94(−.7038)
	U_I	1.4(−.0835)	3.84(+.3309)	3.41(−.486)
	U_w	9.23(−.5658)	8.75(.7545)	1.57(+.2239)
Horizon t = 1	U_C	8.8(−.5417)	3.14(+.2704)	4.43(−.6319)
	U_I	1.7(−.1069)	4.13(+3559)	3.04(−.433)
	U_w	10.6(−.652)	8.33(−.7178)	.03(−.0362)

infinite in some dimension. As the probability level falls the star collapses to the axes. The volume between the star and the ellipsoid represents dZ^* outcomes of a higher probability level. There is therefore a much higher chance of offsetting stochastic shocks on D than C and on either than I (a very low chance), and on G and T rather than W_2. This is some measure of the relative robustness of each Z_i^* to the component shocks. We note that orthogonalising[9] the errors makes little change due to their small covariances.

More straightforward for interpretation[10] is Table 7A.5a whose results confirm Table 7A.1 and the general robustness of Z^*. For subsequent parts of the multiperiod certainty equivalent solution the results will change quite insignificantly as is shown in Table 7A.6 so no further computations were performed.

These results, however, are for Z^* elements individually. None are given for the joint distribution and probabilities of dZ^*. In addition to the effects of individual errors on one variable cancelling, there may be cancelling effects between variables on the welfare index due to some covariance. If two variables are substitutes, the welfare index is largely unaffected even when shocks cause a noticeable increase in one but a suitable decrease in the other. Substitution is also possible between one variable's values over time. Some examples are given in Table 7A.7. We need therefore to complete the analysis of Z^*'s robustness by investigating these substitution possibilities, and to distinguish robustness in Z^* elements from that in the welfare index.

Sensitivity to noncontrollable variables has little meaning in this model. Time has little effect being small in size and coefficient. The constant has more effect as the coefficients are large and it appears in several equations.

Table 7A.3 shows clearly where the specification of the model should be improved. The sensitivity to β_i is largest, particularly when likely variations[11] are considered. These parameters are badly determined in themselves and that equation as a whole is likely to cause large dZ^* values. Only variations in $\alpha_1, \alpha_2, \beta_1, \beta_2, \gamma_0$ and γ_3 tend to have rather little effect. Generally G and I (sometimes W_2) tend to be badly affected and again there is a disparity between relative and absolute disturbances but the absolute ones being of similar size. The chances of cancelling shocks are higher for α_i and γ_i and somewhat lower for I and T. To orthogonalise the parameter shocks is important here reflecting their large covariances. Table 7A.5b[12] shows Z^* clearly less robust particularly in I, T and G. Again the possibility of substitution so the robustness of each Z_i^* to its components varies considerably.

TABLE 7A.7 Welfare losses: multiperiod certainty equivalent solutions taken from earlier tables

Changed X^* caused by:	*1st period*	*2nd period*	*3rd period*	*4th period*	*total*	*% loss*
Table 1, U_C changed by $\pm$ 2 SE	16.68	16.14	14.24	9.71	106.9	8.2
Table 1, U_I changed by $\pm$ 2 SE	14.64	14.09	12.19	7.69	90.38	6.93
Table 1, U_w changed by $\pm$ 2 SE	11.01	10.66	9.46	6.57	69.9	5.36
Table 1(*a*)	19.7	19.64	19.42	17.78	96.09	7.36
Table 1(*b*)	3.96	3.94	3.89	3.56	19.31	1.48
Table 1(*c*)	2.82	2.87	2.77	2.49	14.83	1.13
Table 2, constant	37.6	36.3	31.7	20.72	238.52	18.2
Table 3b (*a*)	7.56	7.53	7.46	6.73	38.01	2.91
Table 3b $\alpha(a)$	33.7	32.4	28.23	18.4	212.8	16.3
Table 3b $\beta(a)$	152.6	146.9	127.1	80.1	942.7	72.3
Table 3b $\gamma(a)$	32.9	31.8	28.2	19.6	208.5	15.9
Table 5a, confidence intervals	6.67	6.64	6.49	5.78	36.12	2.77

XI. CONCLUSIONS

First, we conclude it is important to allow for parameter uncertainty as well as equation error and exogenous information uncertainty if one wants Z^* robust. In this example it is more important. Whether this is best done by first order certainty equivalence as here (Theil (1964)), and allow the optimal revisions of the measured dU in (33) to correct the consequent mistakes, or by an explicit procedure is an open question. This conclusion will extend to other problems.

Secondly, Z^* elements may themselves be robust, but relatively their robustness may vary as can the relative robustness of each Z_i^* to component shocks. Thirdly it is important to account for covariances in the shocks. It is very important to consider covariances (substitution possibilities) between Z_i^* elements. The welfare index may be little affected even when Z_i^* is not robust. This comment applies equally to Z^* obtained by exhaustive simulation but although it may be robust to shocks, the strategy for selecting suitable new policies for simulation may well not be; and there is no guarantee that we in fact get a policy similar to Z^* by simulation and so the welfare index may be highly sensitive in that case.

Fourthly, the specification of (4) (including its functional form if (4) is a quadratic approximation) may be less vital to a robust Z^* than model specification. Obviously this conclusion fades with better specified models and therefore most probably is more important with larger models.

Interpreting ϕ as a regression variance—covariance matrix implies $\max_{i,j}\left|\phi_{ij}\right| \leqslant \max_{i}\left|\phi_{ii}\right| \leqslant \max_{i}\left|Q_{ii}^{-1}\right| \leqslant 1$. Since Z^* is invariant to scalar transformations in Q we can normalise $\max_{i}\left|Q_{ii}^{-1}\right| = 1$.

$$\max_{k}\left|\partial Z_k^*/\partial Q_{ij}\right| \leqslant \left|\tilde{Z}_j^*\right| k, i, j = 1, \ldots, (n+m)T \tag{34}$$

Also if $a = U$, vec N or vec $(\beta\vdots\Gamma)$ then

$$\frac{\partial}{\partial Q_{ij}}\left[\frac{\partial Z^*}{\partial a}\right] = (\phi_{.i})\left[\frac{\partial Z^*}{\partial a}\right]_{j.} \quad j, i = 1, \ldots, (n+m)T \tag{35}$$

or

$$\frac{\partial}{\partial Q}\left[\frac{\partial Z^*}{\partial a}\right] = (\text{vec }\phi)\left[\text{vec}\frac{\partial Z^*}{\partial a}\right] \tag{36}$$

This implies

$$\left|\frac{\partial}{\partial Q_{ij}}\left[\frac{\partial Z^*}{\partial a}\right]\right| < \left|\frac{\partial Z^*}{\partial a}\right|; \text{ and } \frac{\partial(\partial Z_j^*/\partial a)}{\partial Q_{ij}}$$

has the same sign as $\partial Z_j^*/\partial a$ in each element, for all $i, j = 1, \ldots, (n+m)T$. Therefore (a) the change in dZ_i^*, given da, may always be reduced by increasing (decreasing) Q_{ij} if $dZ_j^* = \sum_k \partial Z_j^*/\partial a_k\, da_k$ is positive (negative) – and most effectively by changing Q_{ij}; (b) $\tilde{Z}_i^*$ then rises if $dQ_{ij} \gtrless 0$ and $\tilde{Z}_j^* \gtrless 0$ (falls i $dQ_{ij} \gtrless 0$ and $Z_j^* \gtrless 0$) – and by most if Q_{ii} is used; (c) dZ_j^* and $\hat{Z}_j^*$ are then affected since $Q_{ij} = Q_{ji}$; (d) the changes to dZ_i^* and Z_i^* are less in absolute value than dZ_i^* and dZ_i^* themselves (no sign changes are possible); (e) dQ_{ij} and $-dQ_{ji}$ do not imply the same change to dZ_i^* or $\tilde{Z}_i^*$; (f) increasing $|Q_{ii}|$ implies a reduction in $|\tilde{Z}_i^*|$.

Thus dZ_i^* and $\tilde{Z}_i^*$ may be reduced in absolute size by choice of Q but the scope for doing this may possibly be limited by conflicts between (a) and (b) or (a) and (c). Operating with Q_{ii} for example we can have similar relative dZ_i^* or $\tilde{Z}_i^*$ for each i—it is not necessary to include relative deviations in (4).

Elements of $\phi = (\Omega^{(1)}H - I)Q^{-1}$ in the multiperiod certainty equivalent solution vary as weights vary over time or variables. We can therefore influence the robustness of Z^* (and/or $\tilde{Z}^*$ itself) by choice of Q. When no further progress in reducing those values is possible, then we can pick that Q to make the welfare index robust Therefore even if Z^* is least sensitive to Q, the design of that is central to the reliable use of control methods with a given model.

NOTES

1. In (3) we obviously discard the last $m^*(\ell - 1)$ repeated variables in Y_t^*, the last $n^*(p-1)$ in X_t^* and the associated rows in Π_1 and R_i, $i = 1, \ldots, T$.
2. Symmetry is for convenience—it is not necessary for a general solution. We take Q, q to be adequate to specify the shape of w about its maximum as we want it.
3. This exposition allows coefficient matrices of (2) to vary with time. That can clearly be accommodated in what follows.
4. The Wall-Westcott procedure must have the state vector substituted out.
5. Note Z^* or $\tilde{Z}^*$ here.
6. Let $R^+ = RF$ define F. Then

$$\begin{bmatrix}\theta_{11}^*(i)\\ \theta_{12}^*(i)\end{bmatrix}R^+ = \begin{bmatrix}\theta_{11}^*(i)\\ \theta_{12}^*(i)\end{bmatrix}RF = 0 \quad i = t, \ldots, 1$$

Since by construction of $\theta_{11}{}^*(i)$, $\theta_{12}{}^*(i)$ we have a submatrix of

$$(H_t^* Q_t H_t^{*\prime})^{-1} H_t^* \begin{bmatrix} B & 0 \\ C & 0 \end{bmatrix}_t H_t^{*\prime}$$

7. Increasing any weight on $\tilde{Z}_i$ implies a reduction in $\tilde{Z}_i^*$
8. If regressing K on K^* gives parameter set $\hat{\delta}$, then K on H' gives $K\hat{\delta}$ and the residual variance-covariance matrix is unchanged.
9. Principal components, having the same summed variances, generalised variance but with maximum ordered individual variances to preserve the idea of a 'worst' situation, were used.
10. We assume normally distributed errors.
11. We use the asymptotic variances and asymptotic normality. Small sample corrections would be appropriate. Asymptotic is with respect to the sample size of estimation.
12. (25) is not multivariate normally distributed. To apply the normal distribution here we must apply an ordinary central limit theorem so the results are to be interpreted in an asymptotic sense as $(m+n)T \to \infty$

REFERENCES

(1) Chow, G. C. (1975) *Analysis and control of dynamic linear economic models* (Wiley, New York).

(2) Holbrook, R. S. (1974) A practical method for controlling a large nonlinear stochastic system' *Annals of Econ. and Soc. Meas.*, 3, pp. 155–77.

(3) Leenders, C., and Rothenburg, T. (1964) 'Efficient estimation of simultaneous equation systems', *Econometrica*, 32, pp. 57–67.

(4) Preston, R. S., Klein, L. R., O'Brien, Y. C. and Brown, B. W. (1976). 'Control theory simulations using the Wharton long term annual and industry forecasting model', Wharton Econometric Forecasting Associates Inc. (mimeo).

(5) Rustem B., Vellupillai, K. and Westcott, J. H. (1978) 'A method of respecifying the weighting matrix of a quadratic welfare function' *Automatica*, Sept.

(6) Theil, H. (1964) *Optimal decision rules for government and industry* (North-Holland, Amsterdam).

(7) van den Brogaard, P. and Theil, H. (1959) 'Macrodynamic policy making: an application of strategy and certainty equivalence concepts to the economics of the United States 1933–6', *Metroeconomica.*

(8) Wall, K. D. and Westcott, J. H. 'Policy optimisation studies with a simple control model of the U. K. economy', IFAC/75 Congress, Boston, Mass., Aug. 1975.

Appendix

Sensitivity results for the control problem of van den Brogaard and Theil (1959) using Klein's model 1. The model (FIML) estimates

$$C = \alpha_0 + \alpha_1 + \alpha_2 \Pi_1 + \alpha_3 (w_1 + w_2) + U_c$$
$$w_1 = \gamma_0 + \gamma_1 (Y + T - w_2) + \gamma_2 (Y + T - w_2)_{-1} + \gamma_3 t + U_w$$
$$Y = \Pi + w_1 + w_2$$
$$K = I + K_{-1}$$
$$I = \beta_0 + \beta_1 \Pi + \beta_2 \Pi_{-1} + \beta_3 K_{-1} + U_I$$
$$Y + T = C + I + G$$
$$D = w_1 - 2\Pi$$

C – consumption
I – investment
w_1 – private wage bill
Y – national income
Π – profits
K – capital stock
D – income distribution
w_2 – government wage bill
T – tax bill
G – government expenditure
t – time, 1931 = 1

	α_i	β_i	γ_i
$i = 0$	16.786	17.784	1.599
1	1.02	.231	.42
2	2.225	.546	.164
3	3.8	−.146	.135

Problem horizon $T = 4$; targets C, I, D; instruments w_2, T, G; objective function $A = -I = B; C = 0, q = 0$. Error variances U_C 1.3, U_I 1.42, U_w .61 [Theil (1964)]. Orthogonalised error variances (same ordering) 1.35, 1.9, .075. Estimate variances

[Leenders and Rothenburg (1964)]

α_0 1.645,	α_1 .009,	α_2 .0014,	α_3 .0083
β_0 52.31,	β_1 .0245,	β_2 .0213	β_3 .0012
γ_0 1.285,	γ_1 .0014,	γ_2 .0015,	γ_3 .0008

Orthogonalised estimate variances (same ordering)

2.1175,	.0038,	.00165,	.000292
52.4,	.0277,	.0126,	.000002
.744,	.00002,	.001,	0.0

The errors are assumed approximately normal with zero mean and given variances.

PART III
STABILISATION POLICIES WITH SMALL LINEAR ECONOMETRIC MODELS

8 Control Exercises with a Small Linear Model of the UK Economy

S. Holly, B. Rüstem, J. H. Westcott,
M. B. Zarrop and R. Becker

The use of the methods of optimal control in the formulation of economic policies has been demonstrated in Bray (1975) and Wall and Westcott (1975), as well as in countless other papers. The results reported here are in the same vein as the Bray and Wall and Westcott papers with a linear model, a quadratic objective function and gaussian stochastic disturbances. Rather, however, than go over ground that has already been covered we will focus in this paper on some less central aspects of the control framework. This will provide a complementary study to some more recent work reported in Rustem *et al.* and Zarrop *et al.* (Chapters 6 and 10 of this volume).

The optimal control law will be calculated with the exogenous, uncontrollable, inputs set equal to their mean growth rates. This will be compared with the results of setting the exogenous inputs equal to their historical values, so that in effect the model is subjected to deterministic disturbances. We will also consider how shortening the horizon over which the problem is optimised alters the nature of the control law. The kind of control law that results is tied up with the dynamic properties of the model and this will be demonstrated on a small linear model of the UK economy. In the early stages at least of any realistic exercise in policy formulation the use of control theoretic methods is likely to tell us as much about the properties of the econometric model as about the economic policies governments ought to employ.

Some attention will be paid to ways in which time-varying weighting matrices in the objective function can be used to incorporate constraints upon the frequency with which policy instruments can be used.

Since the conceptual framework has been provided in Westcott *et al.* (Chapter 7 above and references therein), there will be no discussion of the state space form or the nature of the objective function. In Appendix B, however, the direct relationship of the state space form to dynamic multipliers will be spelt out since this has a bearing on the numerical results.

1. ASPECTS OF MODELLING, FORECASTING AND CONTROL

1.1 The Model

The model used in this paper, and outlined in Appendix A, was originally proposed by Bray (1971) though it has gone through several modifications and re-estimations since then. Much of the recent development has been shaped by the demands of the linkup to the London Business School Econometric model which is the subject of Zarrop *et al.* (Chapter 10 of this volume).

The model, in common with most of those used in forecasting and policy simulation is basically an income and expenditure model organised around the system of national accounts, with an income sector that relates disposable income to wages, employment and direct taxation. As the expenditure equations are at constant prices, implicit deflators for gross domestic product and consumption are linked to unit labour costs and import prices. At present there are no monetary variables in the model, which may be a serious omission. Since, however, the contents of this paper are meant to illustrate a particular method, and highlight certain aspects of it, the absence of monetary controls is not too serious. Some progress has been made in incorporating monetary influences but they have not been included in the version of the model used here.

Economic models are employed for a variety of purposes. The best for one purpose will not necessarily be the most appropriate for another. One designed for economic stabilisation may be a very poor forecasting model. But whatever kind of model is suggested it must have some underlying structure which makes economic sense. The determination of the structure of the equations in the model is a critical matter and combines the use of prior economic information, hypothesis testing and correlation techniques where structure is defined to mean the complete functional and stochastic relationship between all endogenous, exogenous and instrument variables. It represents the complete framework of behavioural equations and accounting identities for which a set of parameters has to be obtained by estimation.

1.2 Estimation

The model used in this paper has been estimated in Rational Structural Form (Wall (1976), Wall and Westcott (1974)).

$$\underline{y}(k) = A(z^{-1})\underline{y}(k) + B_1(z^{-1})\underline{u}(k) + B_2(z^{-1})\underline{e}(k) + B_3(z^{-1})\underline{z}(k) + C(z^{-1})\underline{w}(k) \quad (1.1)$$

where

$\underline{y}(k) \triangleq \ell \times 1$ vector of endogenous variables (outputs)
$\underline{u}(k) \triangleq m \times 1$ vector of instruments (controls)
$\underline{e}(k) \triangleq m_1 \times 1$ vector of exogenous variables
$\underline{z}(k) \triangleq m_2 \times 1$ vector of identities
$\underline{w}(k) \triangleq \ell \times 1$ vector of stochastic disturbances

The rational matrix operators $A(z^{-1})$, $B_1(z^{-1})$, $B_2(z^{-1})$, $B_3(z^{-1})$, $C(z^{-1})$ are dimensioned respectively, $\ell \times \ell$, $\ell \times m$, $\ell \times m_1$, $\ell \times m_2$, $\ell \times \ell$.

The elements of the column vector $\underline{w}(k)$ are uncorrelated Gaussian white noise sequences with zero mean and covariance

$$E[\underline{w}(k)\underline{w}^{T}(k+\tau)] = R\delta_{k,k+\tau} \quad (1.2)$$

where R is an $\ell \times \ell$ symmetric matrix.

The determination of structure involves repeated use of single-equation model building as advocated by Box and Jenkins (1970) to find a behavioural equation containing the minimum number of terms that gives a satisfactory account of records.

The technique of structure determination is well developed and can be summarised for each behavioural equation as follows:

(1) pre-whiten and cross-correlate each input with the output to determine how many terms in each of the polynomials of the rational matrices $A(z^{-1})$, $B(z^{-1})$ and $C(z^{-1})$ are justified and furthermore determine the magnitude of pure time delays to be associated with each non-stochastic input on the right-hand side of equation (1.1).
(2) Estimate parameters of all three matrices A, B and C, using the maximum likelihood method.
(3) Arising from the resulting residual correlations redefine structure as necessary and reestimate all parameters.
(4) Repeat steps (2) and (3) until the model passes all diagnostic tests.

Once the structure has been defined it is straightforward to apply a suitable algorithm to obtain parameter estimates. A dynamic generalisation of the method of full information maximum likelihood estimation (FIML) is used to estimate the complete set of equations as in equation (1.1).

On the assumption that the $w(k)$s form a series of independent jointly normal processes a joint probability density function in terms of inputs $\underline{u}(k)$, $\underline{e}(k)$, outputs $\underline{y}(k)$ and computed lagged residuals $\underline{w}(k-j)$ can be obtained.

An interesting advantage which results from this simultaneous estimation of parameters across a set of equations is to yield a more simple and direct set of relationships than would be obtained if the possible coupling between equations were ignored as in single-equation-at-a-time estimation. A worked example showing the advantages of using this simultaneous estimation procedure is given in Wall and Westcott (1974).

It has to be remembered, however, with this estimation method that the underlying assumption is that the parameters being estimated are in reality constants. Much work is being done to extend analysis to cope with the case of time-varying parameters. An approach to this problem is suggested by the Kalman-Bucy filter. Rustem *et al.* (1976) used this filter for an analysis of the time-varying properties of the parameters of an economic model. The filter may be regarded as a Bayesian posterior probability resulting from the modification of the prior probability by a likelihood function. The filter continually updates the conditional probability distributions and when the resulting errors are assumed to be normally distributed the structure of the filter is related in a simple way to the parameters of the distribution functions. It can be employed, therefore, to estimate parameters from data derived from systems the underlying parameters of which are varying with time.

1.3 Linearity of Model

Since many of the economic time series are strongly trended, they are made approximately stationary by expressing most variables in percentage change terms which is almost equivalent to first differences in logarithms. The linear dynamics of the behavioural equations can then be estimated fairly parsimoniously. Thus all of the nonlinear product and quotient terms in the identities that result from deflating variables to constant prices or normalising are reduced to simple sums or differences respectively. However, identities that are originally linear, such as the identity for gross domestic product, are actually nonlinearised when the components of the identity are expressed as percentages. At present these nonlinear identities are linearised about the values of the variables at some time period. For example, the percentage change of total consumption (z_8) in the Prem model is obtained as an identity from the sum of the percentage changes of durable (y_7) and non-durable consumption (y_6). If the upper case denotes the level of the lower case variable, then in percentage changes

$$z_8(k) \triangleq \frac{Y_6(k-1)}{Z_8(k-1)} y_6(k) + \frac{Y_7(k-1)}{Z_8(k-1)} y_7(k) \tag{1.3}$$

where

$$z_8(k) \triangleq \frac{Z_8(k) - Z_8(k-1)}{Z_8(k-1)} \times 100, \quad y_6(k) \triangleq \frac{Y_6(k) - Y_6(k-1)}{Y_6(k-1)} \times 100$$

$$y_7(k) \triangleq \frac{Y_7(k) - Y_7(k-1)}{Y_7(k-1)} \times 100$$

By assuming that the ratios of $Z_8(k)$ to $Y_6(k)$ and $Y_7(k)$ do not change drastically from some initial values, we can replace equation (1.3) by

$$z_8(k) = \alpha y_6(k) + \beta y_7(k)$$

where

$$\alpha \triangleq \frac{Y_6(0)}{Z_8(0)}, \quad \beta \triangleq \frac{Y_7(0)}{Z_8(0)}$$

Needless to say the strength of this assumption will vary with the kind of identity that is being linearised.

2. NUMERICAL RESULTS

The policy instruments in the model are government current expenditure, minimum deposit rate for hire purchase, a weighted exchange rate index, discretionary taxes on expenditure, value added tax, the basic tax rate, aggregated personal allowances and national insurance contributions. It is some reflection on the inherent difficulties of controlling the economy that not one of these supposed instruments is uncontroversial. Either the authorities do not have a sufficiently fine hold over an instrument, for instance government current expenditure, so that it cannot be varied from interval to interval as readily as the optimal control law requires; or else there are severe constraints on the frequency with which certain instruments can be used. Because of the impact that changes in minimum deposit rate have on the output of certain industries there may be good structural and political reasons why it should be used only infrequently, if at all. There may also be serious empirical and theoretical doubts about the extent to which the exchange rate can be used as an instrument of policy, especially in the long run, rather than its being determined endogenously by world and domestic rates of inflation.

The use of optimal control techniques provides no guard against the criticism that the structure of the model which provides the basis for the optimal control calculation is faulty; so that the wrong choice is made when selecting instruments and setting feasible targets. Their use, nevertheless, does have the advantage that the choice of instruments and the establishment of priorities is made quite explicit. Those who doubt the structure can always use the techniques on a model that they can accept, if such a model can be specified and estimated on empirical data.

Before optimal control methods can be used to design economic policies we have to assign values to the desired trajectories for the outputs, $y^d(k)$, and the controls $u^d(k)$. We have to place weights on deviations of actual values from desired values, and fix the trajectories for the exogenous variables. The mix of desired values and weights at this stage must involve a very large element of arbitrariness. Methods have been proposed by which the policy maker can arrive at the set of weights which most nearly approximate his true priorities and the trade-offs between objectives, but since this has already been dealt with in the Rustem *et al.* paper (Chapter 6 of this volume), nothing more will be said here.

Apart from representing the preference ordering of the policy maker, the choice of desired trajectory and the assignment of weights, especially on the controls, provides a useful way of incorporating real institutional constraints on the extent to which certain instruments of policy can be used. For example, the administrative difficulties that arise in changing tax codings as a result of altering tax rates and allowances means that in practice these two controls can be manipulated only once a year at budget time. Since the A and B matrices in the objective function are dated, it is a simple procedure to attach relatively high weights on the controls that we do not wish to change during the first, third and fourth quarters of the calendar year, and set the desired values during these periods to zero. It might also be desirable that an instrument did not change too frequently. For instance, sharp and closely spaced changes in the exchange rate of opposite signs might engender too much uncertainty among exporters who have to enter into commitments now that are binding in the future. After an initial run in which such undesirable changes in the exchange rate were found, an appropriate choice of weights could be used to screen out the offending fluctuation in the exchange rate.

Nevertheless, the choices that are made here are bound to have many elements to which exception can be taken. The point is, however, that we do not claim to be privy to the numerous restrictions that governments will seek to place on the control and output trajectories. We can only indicate how an obvious though limited number can be incorporated.

3.1 Choice of Weights and Desired Trajectories

These are set out in Table 8.1. The control runs were made over the period 1970(1) to 1974(4). Since we are not concerned with the sensitivity of the feedback law to variations in the structure of the cost function, the details of Table 8.1 will remain unchanged between the various runs reported below.

In principle all of the endogenous variables could have been weighted. But to keep the examples straightforward and comprehensive only unemployment, exports, imports and the GDP deflator were weighted. Since the level of unemployment at the initial state was not very high (553,000), or at least with hindsight it does not appear so, it was simply required that unemployment fell gently over the year, with no decline in

TABLE 8.1

1. ENDOGENOUS VARIABLES

desired trajectories

y_1	: %Δ unemployment	
$y_1^d(k)$	$= \{0, -1, -1, -.5, 0 \ldots, 0\}$	$k = 1, 20$
y_{11}	: %Δ exports	
$y_{11}^d(k)$	$= \{1.75, 1.75, \ldots \ldots 1.75\}$,	$k = 1, 20$
y_{12}	: %Δ imports	
$y_{12}^d(k)$	$= \{1.5, 1.5, \ldots \ldots, 1.5\}$,	$k = 1, 20$
y_{18}	: GDP deflator	
$y_{18}^d(k)$	$= \{1.34, 1.34, \ldots, 1.34\}$,	$k = 1, 20$

weights

$a_{1,1}(k) = \{2.0, 2.0, \ldots 2.0\}, k = 1, 20$
$a_{11,11}(k) = \{2.0, 2.0, \ldots 2.0\}, k = 1, 20$
$a_{12,12}(k) = \{2.0, 2.0, \ldots 2.0\}, k = 1, 20$
$a_{18,18}(k) = \{2.0, 2.0, \ldots 2.0\}, k = 1, 20$

TABLE 8.1 *(continued)*

2. POLICY INSTRUMENTS

desired trajectories

$u_1(k)$	: %Δ government current expenditure	
$u_1^d(k)$	= {.437, .437, ...437}	$k = 1, 20$
$u_2(k)$	: Δ minimum deposit rate	
$u_2^d(k)$	= {0, 0, ... 0}	$k = 1, 20$
$u_3(k)$	: weighted exchange rate index	
$u_3^d(k)$	: {0, 0, ... 0}	$k = 1, 20$
$u_4(k)$	: value added tax	
$u_4^d(k)$	: {0, 0, ... 0}	$k = 1, 20$
$u_5(k)$	: cumulated discretionary indirect taxes	
$u_5^d(k)$	= {0, 0, ... 0}	$k = 1, 20$
$u_6(k)$	= basic tax rate	
$u_6^d(k)$	= {0, 0, ... 0}	$k = 1, 20$
$u_7(k)$	: aggregated personal allowances	
$u_7^d(k)$	= {0, 0, ... 0}	$k = 1, 20$
$u_8(k)$	: national insurance contributions	
$u_8^d(k)$	: :{3.17, . . . , 3.17}	$k = 1, 20$

weights

Year j, $j = 1, 5$

	1st Quarter	2nd Quarter	3rd Quarter	4th Quarter
b_{11}	1	1	1	1
b_{22}	3	3	3	3
b_{33}	.01	.01	.01	.01
b_{44}	3	3	3	3
b_{55}	3	3	3	3
b_{66}	15	.01	15	15
b_{77}	15	.01	15	15
b_{88}	1	1	1	1

the first quarter, and remain at the level achieved thereafter. The desired trajectories for exports, imports and prices were set at what was regarded as some reasonable and moderate value. In particular the desired value for prices was set equal to the mean value for the estimation period. All four endogenous variables were weighted equally.

The desired trajectories for the two policy instruments, government current expenditure (u_1) and national insurance contributions (u_8) were set equal to their means over the estimation period. The desired trajectories for the remaining instruments were set equal to zero.

The weights on the controls are chosen so that government expenditure, the exchange rate, national insurance contributions, and the basic tax rate and aggregated allowances (changing only in the second quarter of the year) are used as the main instruments. Indirect taxes and VAT as well as minimum deposit rate were effectively screened out and remained unchanged. When indirect taxes and VAT were given full freedom to vary they tended to change by considerable amounts before they had much impact on the outputs and this was thought unreasonable. Minimum lending rate was also excluded because of the constraints that surround the frequency of its use. Limitations could have been put on the number of times that this instrument was used during the control period but this was not done. It was simply left out.

3.2 CASE I: Optimal Control with Smoothed Exogenous Inputs

Initially, the trajectories for the exogenous variables were set equal to their mean values for the estimation period. This was the approach adopted by both Bray (1975) and Wall and Westcott (1975). Thus

$$\underline{e}^T(k) = [2.143,\ 2.353,\ 2.893,\ 1.039,\ 4.606,\ -.139, 0] \quad k = 1,20$$

The results for the weighted endogenous variables are plotted in Figures 8.3, 4 and 5; those for the controls in Figures 8.1 and 2.

Unemployment fluctuates perversely in the first few quarters before it homes in on the desired trajectory by the sixth quarter. This is achieved by an immediate increase in government expenditure of 3.55 per cent in the first quarter, 1.42 per cent in the next quarter after which government expenditure settles down. In the second quarter basic tax rate is reduced by 9 per cent to 37 per cent in levels, while aggregate allowances go up by almost 7 per cent. Somewhat surprisingly national insurance contributions are not used as a control; u_8 tracks its desired trajectory right through the control period. In the two following years, the tax rate is raised slightly and allowances lowered slightly. In the last two years the opposite happens. These controls are used contracyclically to hold unemployment to its desired path.

When we look at the trajectories of exports, imports and prices it is clear that something peculiar is happening. While imports after the transition period follow a fairly smooth trajectory it is always below the 'desired' trajectory. Since in general we do not want a high rate of increase of imports, this is acceptable. The trajectory for exports is also below its desired level, but since we may want a high rate of increase of exports to maintain a balance of payments surplus this is more serious. It

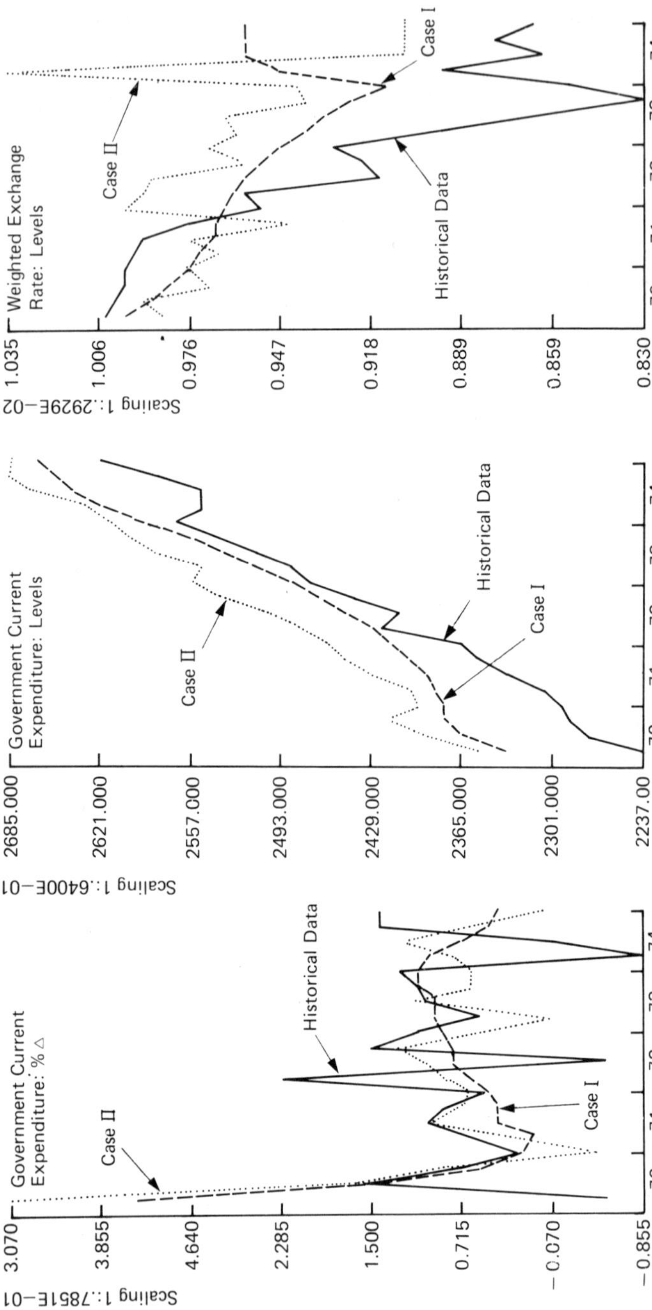
Weighted Exchange
Rate: Levels
Case I
Case II
Historical Data
1.035
1.006
0.976
0.947
0.918
0.889
0.859
0.830
Scaling 1:.2929E−02
70
71
72
73
74
Government Current
Expenditure: Levels
Case II
Case I
Historical Data
2685.000
2621.000
2557.000
2493.000
2429.000
2365.000
2301.000
2237.00
Scaling 1:.6400E−01
Government Current
Expenditure: %△
Case II
Case I
Historical Data
3.070
3.855
4.640
2.285
1.500
0.715
−0.070
−0.855
Scaling 1:.7851E−01

FIG. 8.1

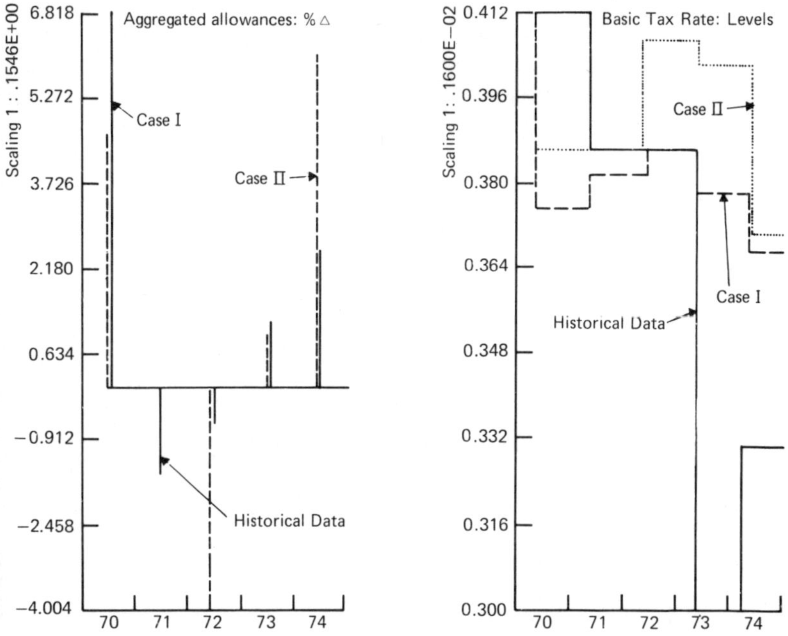

FIG. 8.2

is not difficult to find out why the trade sector does not respond in the way that is expected. While the exchange rate affects exports [y_{11}, in Appendix A] after four quarters with a simple gain, it raises import prices [y_{14}, in Appendix A] gradually over a very long period so that the steady state gain is almost seven, which is plainly wrong. As import prices affect the GDP deflator and this is weighted in the cost function, the optimisation is restraining the growth of exports because of the consequences that the exchange rate depreciation has on prices in the long run (or at least up to five years). The relationship between the state space form and the dynamic multipliers of the model is considered in Appendix B. The dynamic multipliers of the exchange rate for a selected number of variables are reported in Table 8.B.1. Thus while, in this model at least, the exchange rate has a marked impact on prices there is no instrument that can be used to offset its effects. The variations in indirect taxes and VAT that would be needed would be unacceptable. The constraining effect that the long-run consequences of a depreciating

TABLE 8.B.1 Dynamic Multipliers: Selected Inputs/Outputs

	Government expenditure (u_1)				Exchange rate (u_3)						
	y_{16}	y_{12}	y_{18}	y_1	y_{16}	y_{11}	y_{12}	y_{18}	y_{14}	y_1	y_{13}
1	.2489	.1762	−.1189	−.4967	0	0	0	0	0	0	0
2	.2658	.2347	−.1573	−1.198	0	0	0	0	0	0	0
3	.2368	.2222	−.165)	−1.325	.0625	0	.0911	−.0288	−1.000	−.1202	−.6717
4	.2456	.2460	−.1532	−2.321	.0857	0	.1376	−.1334	−1.861	−.3327	−1.249
5	.2466	.2582	−.1396	−2.663	−.0202	−.4779	.0938	−.2045	−2.600	−.3446	−1.757
6	.2441	.2683	−.1194	−2.859	.0022	−.4779	.1196	−.3579	−3.254	−.2945	−2.223
7	.2305	.2642	−.0994	−2.918	.0574	−.4779	.1979	−.5428	−3.780	−.3312	−2.597
8	.2142	.2567	−.0765	−2.863	.0863	−.4779	.2507	−.7496	−4.248	−.464	−2.964
9	.1937	.2420	−.0536	−2.716	.1227	−.4779	.3173	−.9547	4.6506	−.694	−3.282
10	.1732	.2252	−.0314	−2.508	.1565	−.4779	.3811	−1.1605	−4.996	−1.007	−3.565
11	.1540	.2069	−.0126	−2.265	.1935	−.4779	.4508	−1.364	−5.293	−1.378	−3.810
12	.1391	.1913	.0021	−2.017	.2281	−.4779	.5184	−1.568	−5.549	−1.784	−4.027
13	.1294	.1793	.0121	−1.791	.2639	−.4779	.5395	−1.772	−5.768	−2.207	−4.219
14	.1256	.1722	.0176	−1.609	.2981	−.4779	.6604	−1.974	−5.956	−2.634	−4.391
15	.1270	.1700	.0186	−1.484	.330	−.4779	.7306	−2.171	−6.118	−3.052	−4.545
16	.1328	.1724	.0167	−1.423	.3588	−.4779	.7975	−2.366	−6.257	−3.448	−4.682
17	.1414	.1782	.0127	−1.423	.3845	−.4779	.8619	−2.559	−6.376	−3.815	−4.805
18	.1511	.1863	.0082	−1.474	.4080	−.4779	.924	−2.749	−6.479	−4.149	−4.9
19	.1603	.1947	.0048	−1.561	.4297	−.4779	.985	−2.937	−6.567	−4.449	−5.019
20	.1673	.2020	.0035	−1.665	.4500	−.4779	1.043	−3.123	−6.642	−4.721	−5.111

TABLE 8.B.1 (continued)

	Basic tax rate (u_6)					World trade (e_2)		
	y_{18}	y_1	y_{17}	y_6	y_7	y_{16}	y_{11}	y_1
1	.0125	.0522	.8134	−.2176	−.0380	.0520	.1697	−.1040
2	.0093	.1229	.7877	−.2394	−.04172	.0557	.1697	−.2509
3	−.0001	.1780	.7748	−.2093	−.0345	.0496	.1697	−.3823
4	−.0143	.2129	.7507	−.1977	−.0309	.0514	.1697	−.4860
5	−.0295	.2263	.7321	−.1771	−.0256	.0517	.1697	−.5576
6	−.0459	.2210	.7115	−.1625	−.0215	.0511	.1697	−.5937
7	−.0618	.1986	.6960	−.1406	−.0160	.0483	.1697	−.6110
8	−.0776	.1632	.6301	−.1216	−.0111	.0449	.1697	−.5995
9	−.0930	.1136	.6653	−.1012	−.0060	.0406	.1697	−.5689
10	−.1082	.0690	.6502	−.0890	−.0011	.0363	.1697	−.5252
11	−.1228	.0181	.6362	−.0610	.0040	.0322	.1697	−.4743
12	−.1366	−.0309	.6226	−.0420	.0089	.0291	.1697	−.4224
13	−.1496	−.0750	.6092	−.0227	.0138	.0271	.1697	−.3752
14	−.1620	−.1120	.5953	−.0046	.0184	.0263	.1697	−.3370
15	−.1739	−.1407	.5813	.0129	.0228	.0266	.1697	−.3109
16	−.1853	−.1613	.5671	.0299	.0272	.0278	.1697	−.2980
17	−.1965	−.1746	.5528	.0463	.0314	.0296	.1697	−.2979
18	−.2076	−.1822	.5383	.0621	.0355	.0317	.1697	−.3086
19	−.2187	−.1864	.5235	.0775	.0396	.0336	.1697	−.3268
20	−.2301	−.1895	.5087	.0928	.0436	.0350	.1697	−.3487

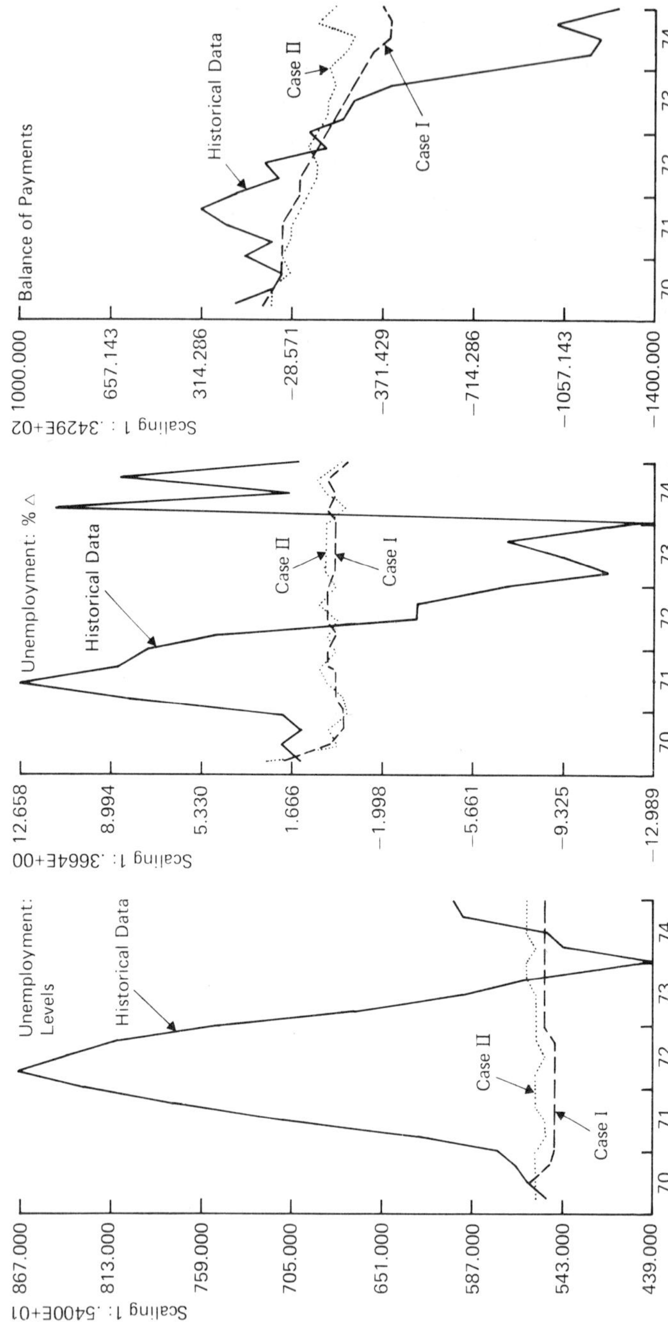
Unemployment: Levels
Historical Data
Case II
Case I
Scaling 1: .5400E+01
867.000
813.000
759.000
705.000
651.000
587.000
543.000
439.000
70
71
72
73
74
Unemployment: % △
Historical Data
Case II
Case I
Scaling 1: .3664E+00
12.658
8.994
5.330
1.666
−1.998
−5.661
−9.325
−12.989
Balance of Payments
Historical Data
Case II
Case I
Scaling 1 : .3429E+02
1000.000
657.143
314.286
−28.571
−371.429
−714.286
−1057.143
−1400.000

FIG. 8.3

exchange rate has for prices, as well as unemployment, can be demonstrated by rerunning the optimal control but weighting only exports among the endogenous variables. The result (Case I.1) for the export trajectory is plotted in Figure 8.4. After four quarters the growth of exports tracks its desired trajectory exactly. This is achieved by a 1.576 per cent depreciation of the exchange rate in each quarter. It is obvious that the trade sector in the model should be reworked. Nevertheless, even if we cannot pretend that these numerical results have any implications for the kinds of policies that governments in the UK could pursue, they do show how the control framework brings out the dynamic features of the model used and exposes any long-run characteristics that might violate theoretical expectations of the long-run behaviour of the real economy.

3.3 CASE II: Optimal Control with Deterministic Disturbances

In this run the exogenous variables were set equal to their historical values over the period 1970(1) to 1974(4). This has the effect of continuously perturbing the optimal trajectories as is shown in Figures 8.1–5. The instruments are also used much more vigorously than when the exogenous variables were smoothed. Of course, in practice the policy maker would not have full information about the exogenous variables, but this example does illustrate that even when we use the principle of certainty equivalence and set the stochastic disturbances to zero, deterministic disturbances in the form of the exogenous variables will produce oscillations in the optimal trajectories. Some of the consequences that this has for control are taken up in Zarrop *et al.* (Chapter 10 below).

3.4 CASE III: Short-run and Long-run Policies

The misspecification in the model and resulting untidiness in the numerical results can be used to illustrate an important point about the choice of period over which the policy-maker decides to optimise his objective function. It is a frequent criticism of economic policy-making that an insufficiently long view is taken of the consequences of certain policies because the electoral process and the uncertainties of forecasting tend to focus attention on the outcome for the economy over the immediate future.

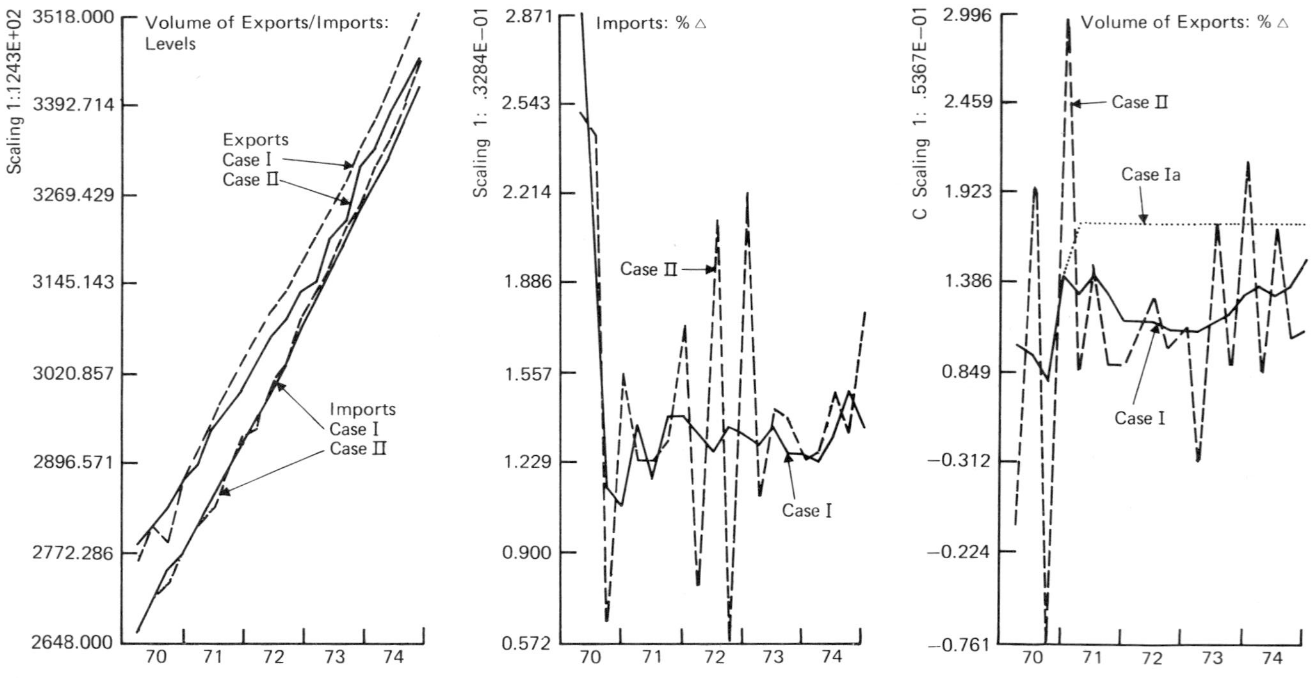
Scaling 1:.1243E+02
3518.000
3392.714
3269.429
3145.143
3020.857
2896.571
2772.286
2648.000
Volume of Exports/Imports:
Levels
Exports
Case I
Case II
Imports
Case I
Case II
70
71
72
73
74
Scaling 1: .3284E−01
2.871
2.543
2.214
1.886
1.557
1.229
0.900
0.572
Imports: % △
Case II
Case I
C Scaling 1: .5367E−01
2.996
2.459
1.923
1.386
0.849
−0.312
−0.224
−0.761
Volume of Exports: % △
Case II
Case Ia
Case I

FIG. 8.4

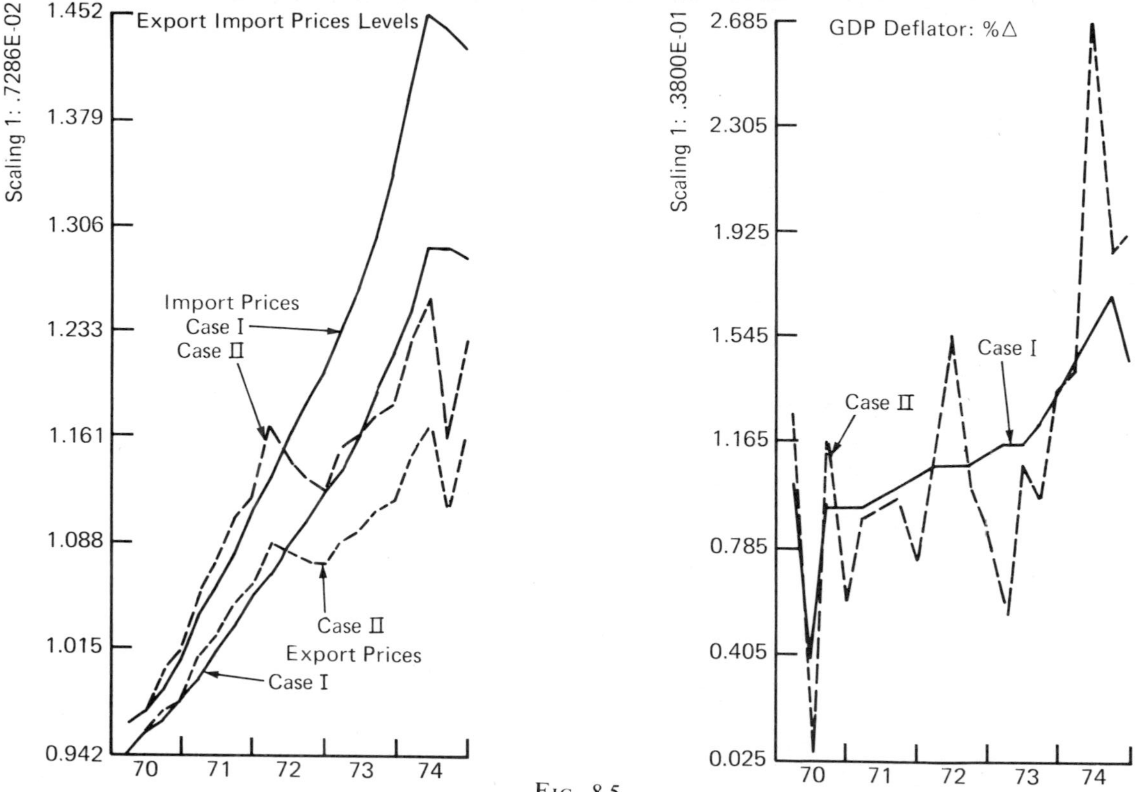
Scaling 1: .7286E-02
1.452
1.379
1.306
1.233
1.161
1.088
1.015
0.942
Export Import Prices Levels
Import Prices
Case I
Case II
Case II
Export Prices
Case I
70
71
72
73
74
Scaling 1: .3800E-01
2.685
2.305
1.925
1.545
1.165
0.785
0.405
0.025
GDP Deflator: %Δ
Case I
Case II
70
71
72
73
74

FIG. 8.5

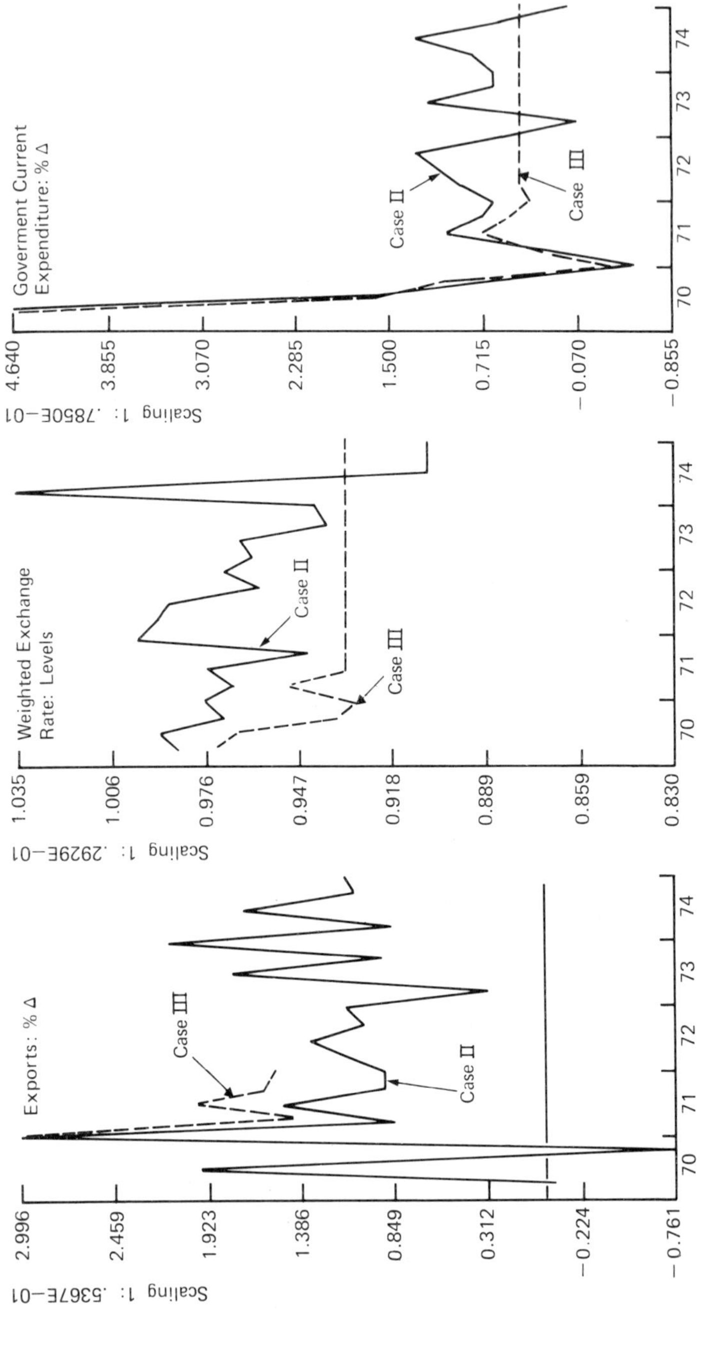
Goverment Current Expenditure: % Δ
Scaling 1: .7850E−01
4.640
3.855
3.070
2.285
1.500
0.715
−0.070
−0.855
Case II
Case III
70
71
72
73
74
Weighted Exchange Rate: Levels
Scaling 1: .2929E−01
1.035
1.006
0.976
0.947
0.918
0.889
0.859
0.830
Case II
Case III
70
71
72
73
74
Exports: % Δ
Scaling 1: .5367E−01
2.996
2.459
1.923
1.386
0.849
0.312
−0.224
−0.761
Case III
Case II
70
71
72
73
74

FIG. 8.6

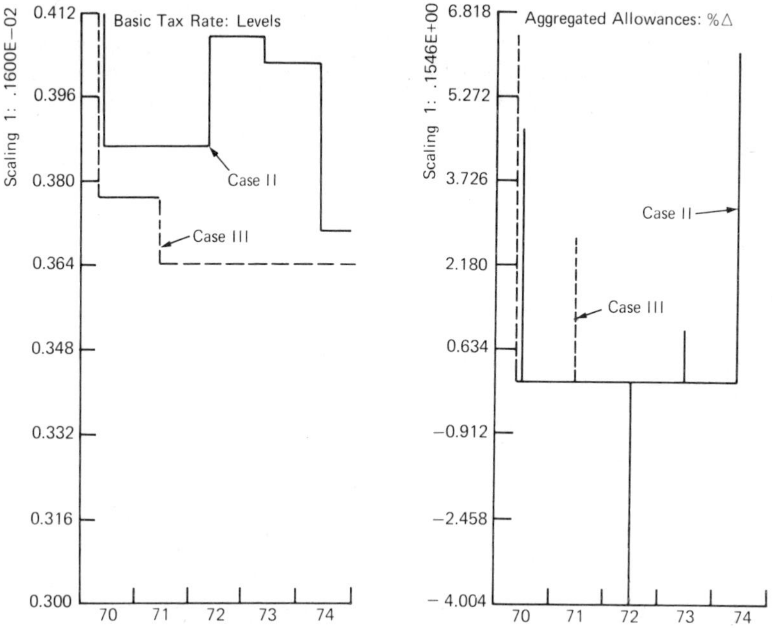

FIG. 8.7

The previous optimal run was repeated but the target endogenous variables were weighted for only the first eight quarters. As can be seen in Figures 8.6 and 7, this results in much more stimulatory action being taken. Though the increase in government expenditure is little different from the previous run, taxes are cut more vigorously. Moreover, the exchange rate depreciates more rapidly and exports are forced up much closer to the desired path. This occurs because the long-run consequences for prices are not being costed by the optimisation criterion. It must be repeated that this particular example does not reflect any realistic trade-off over time. Nevertheless, it may well be that there are real trade-offs over time between objectives and if they can be embodied in the structure of the model the proper choice of the planning horizon will reveal them. Policy-makers will then have a clearer idea of the long-term costs of pursuing short-term objectives.

CONCLUSIONS

The introduction of sophisticated methods from the field of control, though a major advance on those used at present, does not dislodge the economic model from its central place in the formulation of economic policies. Control methods provide powerful tools when they are used wisely but they should not distract attention from the theoretical specification, estimation and diagnosis of economic models which have a very important bearing on the eventual choice of instruments and targets and on the kinds of trajectories that are believed feasible.

REFERENCES

Box, G. E. P. and Jenkins, G. M. (1970) 'Time Series Analysis, Forecasting and Control' (Holden Day).

Bray, J. (1971) 'Proposed Structure for a Simple Control Model of the U.K. Economy', PREM Discussion Paper no. 1.

Bray, J. (1975) 'Optimal Control of a Noisy Economy with the U.K. as an Example', *J. R. Stat. Soc.* (A), 138.

Preston, A. J., and Wall, K. D. (1973), 'Some Aspects of the Use of State Space models in Econometrics', PREM Discussion Paper no. 5.

Rüstem, B. *et al.* (1976), 'Recursive Parameter Estimation using the Kalman Filter: An application to Analyse Time-Varying Model Parameters and Structural Change', Proc. European Meeting of the Econometric Society, Helsinki, 23–27 Aug.

Rüstem, B. *et al.* (1977), 'Iterative Respecification of a Quadratic Objective Function' Chapter 6 of this volume.

Wall, K. D. (1976), 'FIML Estimation of Rational Distributed Lag Structure Form Models', PREM Discussion Paper no. 11.

Wall, K. D. and Westcott, J. H. (1974), 'Macroeconomic Modelling for Control', *IEEE Trans. On Aut. Control*, AC–19, no. 6, Dec.

Wall, K. D. and Westcott, J. H. (1975), 'Policy optimisation studies with a simple control model of the U.K. Economy', IFAC/75 Congress, Boston, Mass., 24–30 Aug.

Westcott, J. H., *et al.* (1977), 'A Control Theory Framework for Policy Analysis' (Chapter 7 above).

Zarrop, M. E., *et al.* (1977), 'Control of the LBS Econometric Model via a Control Model' (Chapter 10 below).

Appendix A The Prem Model

A.1.1 Employment-Unemployment

The rate of growth of unemployment (y_1), apart from a trend term subsumed in the mean, is explained solely by the rate of growth of gross domestic product (y_{16}). The dynamics of the transfer function, y_1/y_{16}, indicate that a sustained 1 per cent rise in the growth rate of GDP will eventually reduce unemployment by 11 per cent.

Once unemployment is determined the level of employment could be obtained via an identity from an exogenously determined labour supply. An alternative procedure, and one adopted with the Prem model, is to explain the growth rate of employment as well. The result is a somewhat unconventional equation that explains the rate of growth of employment (y_4) by the rate of growth of earnings (y_9). The transfer function, y_4/y_9, is well determined, although there is a pure delay of five quarters. Wage inflation reduces employment only after fifteen months.

A.1.2 Stockbuilding

Stockbuilding (y_3), in contrast to practically all the other variables in the model, is normalised with respect to GDP (y_{16}), rather than itself. This is done because stockbuilding is already derived as the difference between two stock levels, rather than as a flow, as the other components of expenditure are. The rate of growth of GDP is used as an explanatory variable. The transfer function, y_3/y_{16}, has a simple first order autoregressive structure.

A.1.3 Investment

Gross domestic fixed capital formation is disaggregated into private fixed non-residential investment (y_2), private residential investment (y_3), and public sector investment which is exogenous. The investment equations in the Prem model are probably its weakest point, apart from the trade sector, and possibly a greater degree of disaggregation and a broader range of explanatory variables would be worth while.

Private fixed non-residential investment is explained by a second order autoregressive structure on GDP superimposed on a pure delay of three quarters. The lag structure is quite plausible. A damping factor of .82 provides a distributed effect of output on investment that is spread over three years or so, which is in accordance with other studies that find that the lag between output and investment is very long.

Private residential investment was the source of even greater problems. Ideally, this form of investment should be explained as a part of a model of building society behaviour. This, however, would have meant increasing the number of behavioural equations in the model by about five. Rather than adopt this course, an alternative hypothesis was used. It is postulated that residential construction receives residual supplies of factor resources from other sectors of the economy. Thus private non-residential investment was used as an explanatory variable. The final result was a transfer function, y_5/y_2, with a distributed lag on a three quarter delay.

A.1.4 Consumption

Consumption is disaggregated into durable and non-durable consumption. For non-durable consumption the standard explanatory variables, the real value of current grants to the personal sector, and real disposable income net of real value of current grants, are used. Both variables have simple transfer function gains with no dynamics or delays. The second order dynamics of the error term, however, suggest that there is some process of adjustment in non-durable consumption that is not accounted for by disposable income.

Consumer durable expenditure is a function of real disposable income, net of the real value of current grants, and the first difference of the minimum deposit rate for hire purchase. Again the transfer functions have simple gains while there are second order dynamics on the errors. Both equations were finally estimated simultaneously.

A.1.5 Wages and Profits

The wage or earnings rate (y_9) has been defined as total wages and salaries plus employers' national insurance contributions deflated by the total number employed times the average weekly hours worked (e_6). The rate of growth of the wage rate is explained by the rate of growth of the implicit deflator for consumption (z_8), a proxy for productivity (z_7)—

defined as gross domestic product over total employment—and the rate of change of unemployment (y_1). The most notable aspect of this equation is the presence of unemployment as an explanatory variable with a pure delay of one quarter. Its coefficient is marginally significant, but the variable has been retained since otherwise the prices sector become indeterminate.

The rate of growth of gross trading profits (Y_{10}) is dependent on the rate of growth of consumer prices (z_8), real disposable income net of government grants (z_1), and the wage rate (y_9), but without any dynamics or pure delays. The error term is fully modelled by a first order autocorrelation. Both wages and profits are estimated simultaneously.

A.1.6 Imports-Exports

Relatively simple aggregate functions are used, even though disaggregation is probably desirable.

The explanatory variables used in the export equation are the rate of growth of world exports (e_2), a weighted exchange rate (u_3) and a dummy variable (e_7) to take account of the dock strikes of 1967 and 1972. The differencing operation prior to estimation tended to amplify the effects of the strike on the time series for exports. Failure to include the dummy variable produced entirely misleading results for the y_{11}/u_3 transfer function because both strikes coincided with, or followed immediately after, changes in the exchange rate. Once this spurious correlation had been filtered out it was possible to detect a simple gain with a delay of four quarters. The result, however, is not particularly significant statistically.

The import function proved very difficult to fit satisfactorily. The rate of growth of imports (y_{12}) is related simply to total final expenditure (z_9). No price effects could be detected. An attempt was made to separate out stockbuilding as having a distinct effect on imports but this was unsuccessful. The finally estimated equation has a simple gain on total final expenditure, no dynamics and a first order autocorrelated noise term.

A.1.7 Export and Import Prices

The rate of growth of export prices (y_{13}) is determined by import prices (y_{14}) in the current period and by unit labour costs (z_{11}) delayed two quarters. The error process is adequately described by a first order autoregression.

The rate of growth of import prices (y_{14}) is determined by the exchange rate (u_3). Attempts to detect any effects of world commodity prices on import prices proved unsuccessful. The y_{14}/u_3 transfer function, though well determined statistically, is very suspect. The steady state gain is clearly implausible, and the damping factor suggests that the effects of a unit change in the exchange rate are spread over a considerable period of time. The dynamics of the error process with a first order denominator suggest some fairly fundamental mis-specification.

A.1.8 Factor Cost Adjustment

Factor cost adjustment (y_{15}) is explained by simple gains on total consumption (z_8), exports (y_{11}), total investment (z_{10}) and a first order autocorrelation.

A.1.9 A Dynamic Identity for GDP

GDP (y_{16}) is included among the behavioural equations so that it can be assigned a weight in the quadratic welfare function if so desired. Identities are not weighted since they are substituted out before the state space form is derived. The noise is set to an arbitrarily small number.

A.1.10 Taxes on Personal Income

This is usually regarded as being one of the more difficult sectors to model because of the profusion of tax rates, allowances and non-linearities. Nevertheless, it is important that personal taxes are included in the model, not only so that disposable income can be obtained, but also because for control purposes tax rates may be important policy instruments. The difficulties of modelling have been exacerbated by the linkup with LBS model which has a very sophisticated disaggregated tax sector.

One way in which the personal tax yield series can be modelled is by taking explicit account of changes in tax instruments, i.e. controls, so that each instrument's effect is reflected in individual regression coefficients. If there were only one tax rate, and we had data on the personal income and total allowances of each individual taxpayer, and there was no measurement error, it would be possible to determine the total personal tax yield by aggregating over all taxpayers. In practice there is more than one tax rate which can vary with the level of income and its source (such as whether income is earned or unearned). There are also a number of inconsistencies between Inland Revenue data and the National Income series. The true model is, therefore, nonlinear and subject to an unknown error process.

It would be possible to refine the data to take into account the inconsistencies between the Inland Revenue and the National Income series but this would make the model too large. Instead, as a simple, but brutal, expedient, the data inconsistencies were ignored; only one tax rate, the basic tax rate, was used, and all major allowances were aggregated.

The rate of growth of personal taxes (y_{17}) was then regressed on personal income (z_4), aggregated allowances (u_6) and the basic tax rate (u_7). The deviation of the estimated coefficients from their *a priori* values is some indication of how good a linear approximation this equation is. The fourth order autocorrelation in the error is probably due to the fact that instruments tend only to be changed at budget time, once a year.

A.1.11 Implicit Deflators

a. GDP Deflator The rate of growth of the GDP deflator (y_{18}) is explained by the rate of growth of unit labour costs (z_{11}) defined as total wages and salaries plus employers' national insurance contributions divided by total employment. The transfer function, y_{18}/z_{11}, gives a straightforward geometric lag profile. The error process is adequately modelled by a first order autocorrelation term.

b. Durable Consumption Deflator As with the GDP deflator, the rate of growth of the durable consumption deflator (y_8) is explained by unit labour costs. The damping factor in the denominator is larger, however, so that the lag is longer.

c. Consumption Deflator The rate of growth of the deflator for total consumption (z_{12}) is explained by a simple gain on unit labour costs delayed one quarter, as well as by the difference between the rate of growth of the GDP deflator (y_{18}) and import prices (y_{14}) and the difference between the rate of growth of the GDP deflator and the 'implicit deflator' for factor cost adjustment (z_{16}) which reflects the impact of subsidies and indirect taxes on consumer prices. The coefficients on these two latter variables are set equal to unity prior to estimation.

d. Non-Durable Consumption Deflator The rate of growth of the deflator for non-durable consumption (z_6) is determined as an identity from the deflators for total consumption and durable consumption.

A.1.12 Taxes on Expenditure

Taxes on expenditure are needed in the model to capture the effects of indirect taxes. The relationship is estimated as an aggregate function but the result is not very satisfactory. The rate of growth of taxes on expenditure (z_{14}) is explained by a distributed lag on the rate of growth of total consumption at current prices (z_{17}) along with the rate of VAT times the regulator (u_5) and discretionary changes in indirect taxes (u_4). The coefficients on the latter two variables are fixed prior to estimation to be unity.

A.2 THE MODEL EQUATIONS

A.2.1 Behavioural Equations

1. Rate of growth of unemployment

$$y_1(k) = \frac{\underset{(.981)}{-1.995}}{(1 - \underset{(.315)}{1.345}B + \underset{(.265)}{.521}B^2)} y_{16}(k) + (1 + \underset{(.094)}{.367}B)\, w_1(k) + 9.117$$

$$R^2 = .54$$

2. Rate of growth of private fixed non-residential investment

$$y_2(k) = \frac{(\underset{(.241)}{.720} - \underset{(.248)}{.622}B)}{(1 - \underset{(.072)}{1.643}B + \underset{(.073)}{.820}B^2)} y_{16}(k-3) + (1 - \underset{(.132)}{.640}B)\, w_2(k) + .809$$

$$R^2 = .28$$

3. Growth in stockbuilding as percentage of GDP

$$y_3(k) = \frac{\underset{(.067)}{.415}}{(1 - \underset{(.041)}{.876}B)} y_{16}(k) + \frac{w_3(k)}{(1 - \underset{(.119)}{.245}B)} 1.328$$

$$R^2 = .52$$

4. Growth rate of employment

$$y_4(k) = \frac{\underset{-.085}{(.028)}}{(1 - \underset{(.121)}{.428}B + \underset{(.137)}{.652}B^2)} y_9(k-5) + \frac{w_4(k)}{(1 - \underset{(.109)}{.488}B^2)} + .892$$

$R^2 = .25$

1 ~ 4 estimated simultaneously

Covariance Matrix

	y_1	y_2	y_3	y_4
y_1	30.32			
y_2	−.2459	11.95		
y_3	.2099	−.4230	.4440	
y_4	−.7714	.2897	−.0552	.1159

5. Growth rate for private residential investment

$$y_5(k) = \frac{\underset{-.3926}{(.1523)}}{(1 - \underset{(.1704)}{.8272}B)} y_2(k-3) + (1 + \overset{(.1259)}{.5899}B^3)\, w_5(k) + 3.975$$

$R^2 = .345 \quad \sigma^2 = 12.25$

6. Growth rate for consumer non-durable expenditure

$$y_6(k) = \overset{(.059)}{.347}z_1(k) + \overset{(.022)}{.073}z_{13}(k) + \frac{w_6(k)}{(1 + \underset{(.210)}{.684}B + \underset{(.17)}{.263}B^2)} + .314$$

$R^2 = .43$

7. Growth rate for consumer durable expenditure

$$y_7(k) = \overset{(.419)}{1.897}z_1(k) - \overset{(.168)}{.836}u_2(k) + \frac{w_7(k)}{(1 + \underset{(.126)}{.481}B + \underset{(.127)}{.596}B^2)} + .695$$

$R^2 = .52$

6 ~ 7 estimated simultaneously

Covariance Matrix

	y_6	y_7
y_6	.5044	
y_7	1.72	26.75

8. Growth rate of implicit deflator for durable consumption

$$y_8(k) = \frac{\underset{(.0457)}{.1849}}{[1 - \underset{(.1447)}{1.5038}B + \underset{(.1109)}{.8081}B^2]} z_{11}(k) + w_8(k) - .169$$

$R^2 = .494 \quad \sigma^2 = .9638$

9. Growth rate of hourly wage rate

$$y_9(k) = \underset{(.222)}{.737} z_{12}(k-1) + \underset{(.084)}{.214} z_7(k-2) - \underset{(.016)}{.028} y_1(k-1) + (1 + \underset{(.133)}{.418} B^3) w_9(k) + 1.302$$

$R^2 = .32$

10. Growth rate of company gross trading profits

$$y_{10}(k) = \underset{(.600)}{3.012} z_{12}(k) + \underset{(.354)}{3.078} z_1(k) - \underset{(.348)}{3.358} y_9(k) + (1 - \underset{(.125)}{.359} B) w_{10}(k) + 3.254$$

$R^2 = .63$

9 ~ 10 estimated simultaneously

Covariance Matrix

	y_9	y_{10}
y_9	.9266	
y_{10}	.3056	9.22

11. Growth rate of export of goods and services

$$y_{11}(k) = \underset{(1.429)}{11.538}\, e_7(k) - \underset{(.2870)}{.4776}\, u_3(k-4) + \underset{(.0653)}{.1697}\, e_2(k) + (1 - \underset{(.1647)}{.3951} B) w_{11}(k) + .561$$

$R^2 = .694 \quad \sigma^2 = 5.579$

12. Growth rate of imports of goods and services

$$y_{12}(k) = \underset{(.2002)}{1.3909}\, z_9(k) + (1 - \underset{(.172)}{.435})w_{12}(k) + .081$$

$R^2 = .463 \quad \sigma^2 = 5.058$

13. Growth rate of export price index

$$y_{13}(k) = \underset{(.081)}{.671}\, y_{14}(k) + \underset{(.059)}{.212}\, z_{11}(k-2) + (1 - \underset{(.114)}{.459}B)w_{13}(k) + .044$$

$R^2 = .496 \quad \sigma^2 = 1.466$

14. Growth rate of import price index

$$y_{14}(k) = \frac{\overset{(.1611)}{-1.001}}{\underset{(.0436)}{(1 - .8591B)}} u_3(k-2) + \frac{\overset{(.2165)}{[1 - .5704B]}}{\underset{(.1789)}{[1 - .7983B]}} w_{14}(k) - .332$$

$R^2 = .639 \quad \sigma^2 = 3.73$

15. Growth rate of factor cost adjustment

$$y_{15}(k) = \underset{(.158)}{1.428}\, z_8(k) + \underset{(.069)}{.099}\, y_{11}(k) + \underset{(.056)}{.111}\, z_{10}(k) + (1 - \underset{(.092)}{.577}B)w_{15}(k) - .365$$

$R^2 = .608 \quad \sigma^2 = .01$

16. GDP (dynamic identity treated as a behavioural equation with zero noise)

$$y_{16}(k) = (1 - 1.0B)y_3(k) + .763 z_8(k) + .293 y_{11}(k) + .215 z_{10}(k) - .300 y_{12}(k) + .216 e_4(k) - .191 y_{15}(k)$$

17. Growth rate of personal tax yield

$$y_{17}(k) = \underset{(.2149)}{1.8687} z_4(k) + \underset{(.1743)}{.8134} u_6(k) - \underset{(.113)}{.6144} u_7(k) + (1 - \underset{(.1418)}{.4244}B + \underset{(.214)}{.5697}B^4)w_{17}(k) + .132$$

$R^2 = .684 \quad \sigma^2 = 6.78$

18. Growth rate of implicit deflator for GDP

$$y_{18}(k) = \underset{(.120)}{\frac{\overset{(.0920)}{.4776}}{[1-.4986B]}} z_{11}(k) + [1 - \overset{(.098)}{.375}B] w_{18}(k) - .113$$

$R^2 = .5285$ $\sigma^2 = 1.078$

19. Growth rate of *transformed* implicit deflator for total consumption

$$y_{19}(k) = \overset{(.192)}{.498} z_{11}(k-1) + [1 + \overset{(.151)}{.396}B] w_{19}(k) + .316$$

$R^2 = .2057$ $\sigma^2 = 3.40$

20. Growth rate of *transformed* taxes on expenditure

$$y_{20}(k) = \underset{(.1128)}{\frac{\overset{(.3850)}{1.070}}{[1-.8258B]}} z_{17}(k) + w_{20}(k) - 9.190$$

$R^2 = .1825$ $\sigma^2 = 3.53$

A.2.2. Identities (levels)

Z_1 : Real personal disposable income excluding Government grants to the personal sector

$$Z_1(k) = Z_5(k)/Z_{12}(k)$$

Z_2 : Total wages and salaries plus employers' national insurance contributions

$$Z_2(k) = Y_9(k) * E_6(k) * Y_4(k)$$

Z_3 : Deductions from taxable personal income

$$Z_3(k) = Y_{17}(k) + U_8(k)$$

Z_4 : Personal income

$$Z_4(k) = Z_2(k) + E_3(k) + E_1(k)$$

Z_5 : Personal disposable income excluding Government grants

$$Z_5(k) = Z_4(k) - Z_3(k) - E_3(k)$$

Z_6 : Implicit deflator for non-durable consumption

$$Z_6(k) = [Z_8(k) * Z_{12}(k) - Y_7(k) * Y_8(k)]/Y_6(k)$$

Z_7 : Productivity proxy

$$Z_7(k) = Y_{16}/Y_4$$

Z_8 : Total consumption

$$Z_8(k) = Y_6(k) + Y_7(k)$$

Z_9 : Total final sales

$$Z_9(k) = Z_8(k) + Z_{10}(k) + Y_{11}(k) + U_1(k)$$

Z_{10} : Total fixed investment

$$Z_{10}(k) = Y_2(k) + Y_5(k) + E_4(k)$$

Z_{11} : Unit labour costs

$$Z_{11}(k) = Z_2(k)/Y_{16}(k)$$

Z_{12} : Implicit deflator for total consumption

$$Z_{12}(k) = Y_{19}(k) + .857\left[\frac{Y_{15}(k)}{Z_8(k)}\{Z_{16}(k) - Y_{18}(k)\}\right] + .652\left[\frac{Y_{12}(k)}{Z_8(k)}\{Y_{14}(k) - Y_{18}(k)\}\right]$$

Z_{13} : Real value of Government grants to the personal sector

$$Z_{13}(k) = E_3(k)/Z_{12}(k)$$

Z_{14} : Taxes on expenditure

$$Z_{14}(k) = [Y_{20}(k) + U_5(k)]/U_4(k)$$

Z_{15} : Taxes on expenditure plus subsidies

$$Z_{15}(k) = Z_{14}(k) + E_5(k)$$

Z_{16} : Implicit deflator for factor cost adjustment

$$Z_{16}(k) = Z_{15}(k)/Y_{15}(k)$$

Z_{17} : Total consumption at current prices

$$Z_{17}(k) = Z_8(k) * Z_{12}(k)$$

A.2.3 Identities (growth rates linearised about 1973 (4))

$$z_1(k) = z_5(k) - z_{12}(k)$$
$$z_2(k) = y_9(k) + e_6(k) + y_4(k)$$
$$z_3(k) = .664y_{17}(k) + .336u_8(k)$$
$$z_4(k) = .639z_2(k) + .102e_3(k) + .208e_1(k)$$
$$z_5(k) = 1.411z_4(k) - .267z_3(k) - .144e_3(k)$$
$$z_6(k) = 1.1006z_8(k) + 1.1006z_{12}(k) - .1006y_7(k) - .1006y_8(k)$$
$$z_7(k) = y_{16} - y_4(k)$$
$$z_8(k) = .903y_6(k) + .097y_7(k)$$
$$z_9(k) = .513z_8(k) + .144z_{10}(k) + .197y_{11}(k) + .145u_1(k)$$
$$z_{10}(k) = .502y_2(k) + .107y_5(k) + .391e_4(k)$$
$$z_{11}(k) = z_2(k) - y_{16}(k)$$
$$z_{12}(k) = y_{19}(k) + .2166y_{14}(k) + .2610z_{15}(k) - .4776y_{18}$$
$$z_{13}(k) = e_3(k) - z_{12}(k)$$
$$z_{14}(k) = y_{20}(k) + .1u_5(k) + .036u_4(k)$$
$$z_{15}(k) = .8496z_{14}(k) + .1504e_5(k)$$
$$z_{16}(k) = z_{15} - y_{15}$$
$$z_{17}(k) = z_{12} + z_8$$

A.2.4 Mnemonics

Endogenous Variables

y_1 : Rate of growth of unemployment
y_2 : Rate of growth of private fixed non-residential investment
y_3 : Growth of stockbuilding as a percentage of GDP
y_4 : Rate of growth of employment
y_5 : Rate of growth of private residential investment
y_6 : Rate of growth of consumer non-durable expenditure
y_7 : Rate of growth of consumer durable expenditure
y_8 : Rate of growth of implicit deflator for durable consumption
y_9 : Rate of growth of hourly wage rate
y_{10} : Rate of growth of company gross trading profits
y_{11} : Rate of growth of exports of goods and services
y_{12} : Rate of growth of imports of goods and services
y_{13} : Rate of growth of the implicit deflator for exports of goods and services
y_{14} : Rate of growth of the implicit deflator for imports of goods and services
y_{15} : Rate of growth of factor cost adjustment
y_{16} : Rate of growth of gross domestic product (dynamic identity)

y_{17} : Rate of growth of direct personal tax
y_{18} : Rate of growth of implicit deflator for gross domestic product
y_{19} : Rate of growth of implicit deflator for consumption *transformed*
y_{20} : Rate of growth of taxes on expenditure *transformed*

Exogenous Variables

e_1 : Rate of growth in 'other' persona income
e_2 : Rate of growth in world exports
e_3 : Rate of growth in Government current grants to personal sector
e_4 : Rate of growth in Government capital expenditure
e_5 : Subsidies
e_6 : Average hours worked
e_7 : Dummy for dock strikes of 1967 (4) and 1972 (3)

Controls

u_1 : Rate of growth in Government current expenditure
u_2 : Difference in minimum deposit rate for hire purchase
u_3 : Rate of growth of weighted exchange rate index
u_4 : Rate of growth of cumulated indirect taxes
u_5 : VAT and the regulator
u_6 : Basic tax rate
u_7 : Aggregated allowances
u_8 : National insurance contributions

Source of Data

All data series have been provided by the London Business School.

Appendix B The Markov Matrices of the Prem Model

Information about the overall dynamic relationships between inputs and outputs of the model is not easily obtained from the study of individual equations especially when they are written in rational structural form.

One of the advantages of the state space representation of econometric models is the ease with which information about the dynamic behaviour of a model can be obtained and the contribution of each part of the system, both deterministic and stochastic, revealed.

Given the polynomial structural form of an econometric model (Preston and Wall (1973))

$$\hat{A}(z^{-1})\underline{y}(k) = \hat{B}(z^{-1})\underline{u}(k) + \hat{C}(z^{-1})\underline{e}(k) + \hat{D}(z^{-1})\underline{w}(k) \qquad \text{(B.1)}$$

$$w(k) \sim N(0, \Omega)$$

where

$y(k)$ is an $\ell \times 1$ vector of endogenous variables
$u(k)$ is an m $\times$ 1 vector of controls
$e(k)$ is an $m_1 \times 1$ vector of exogenous variables
$w(k)$ is an $\ell \times 1$ vector of disturbances.

$\hat{A}, \hat{B}, \hat{C}, \hat{D}$ are $\ell \times \ell$, $\ell \times$ m, $\ell \times m_1$, $\ell \times \ell$ polynomial matrices in the lag operator z^{-1}, and defined as

$$\hat{A}(z^{-1}) = \sum_{i=0}^{p} \hat{A}_i z^{-i}; \qquad \hat{B}(z^{-1}) = \sum_{i=0}^{q} \hat{B}_i z^{-i}; \qquad \hat{C}(z^{-1}) = \sum_{i=0}^{r} \hat{C}_i z^{-i};$$

$$\hat{D}(z^{-1}) = \sum_{i=0}^{s} \hat{D}_i z^{-i}.$$

For $\hat{A}_O$ nonsingular, the reduced form removes the instantaneous coupling between the outputs. Thus

$$A(z^{-1})\underline{y}(k) = B(z^{-1})\underline{u}(k) + C(z^{-1})\underline{e}(k) + D(z^{-1})\underline{w}(k) \qquad \text{(B.2)}$$

where

$$A = \hat{A}_0^{-1}\hat{A};\ B = \hat{A}_0^{-1}\hat{B};\ C = \hat{A}_0^{-1}\hat{C};\ D = \hat{A}_0^{-1}\hat{D}.$$

If the roots of $A(z^{-1})$ lie outside the unit circle, $A(z^{-1})$ may be inverted to give the final form representation

$$\underline{y}(k) = \Pi(z^{-1})\underline{u}(k) + \Lambda(z^{-1})\underline{e}(k) + \Gamma\underline{w}(k) \tag{B.3}$$

where

$$\Pi(z^{-1}) = A^{-1}(z^{-1})B(z^{-1})$$
$$\Lambda(z^{-1}) = A^{-1}(z^{-1})C(z^{-1})$$
$$\Gamma(z^{-1}) = A^{-1}(z^{-1})D(z^{-1}).$$

The final form is equivalent to the input-output representation that control engineers use.

For control purposes, the final form must be translated into a representation that involves only first order difference equations, the state space form. The system of equations

$$\underline{x}(k+1) = F\underline{x}(k) + G_u\underline{u}(k) + G_e\underline{e}(k) + G_w\underline{w}(k) \tag{B.4.a}$$

$$\underline{y}(k) = H\underline{x}(k) + D_u\underline{u}(k) + D_e\underline{e}(k) + D_w\underline{w}(k) \tag{B.4.b}$$

is called the state space representation of equation (A.3).

The task of getting from (B.3) to (B.4) is denoted a *realisation*, and amounts to determining the set of constant matrices $\{F, G_u, G_e, G_w, H, D_u, D_e, D_w\}$ such that the mapping $\{\underline{u}(k), \underline{e}(k), \underline{w}(k)\} \rightarrow \{\underline{y}(k)\}$ defined by (B.3) is completely described by (B.4).

Since, using the forward shift operator z, (B.4.a) can be written

$$[zI - F]\underline{x}(k) = G_u\underline{u}(k) + G_e\underline{e}(k) + G_w\underline{w}(k) \tag{B.5}$$

substitution into (B.4.b) gives

$$\begin{aligned}\underline{y}(k) = {} & [H[I - z^{-1}F]^{-1}G_u + D_u]\underline{u}(k-1) + [H[I - z^{-1}F]^{-1}G_e \\ & + D_e]\underline{e}(k-1) + [H[I - z^{-1}F]^{-1}G_w + D_w]\underline{w}(k-1)\end{aligned} \tag{B.6}$$

so that from (B.3)

$$\begin{aligned}\Pi(z^{-1}) &= Hz^{-1}[I - z^{-1}F]^{-1}G_u + D_u \\ \Lambda(z^{-1}) &= Hz^{-1}[I - z^{-1}F]^{-1}G_e + D_e \\ \Gamma(z^{-1}) &= Hz^{-1}[I - z^{-1}F]^{-1}G_w + D_w.\end{aligned} \tag{B.7}$$

Taking power series expansions

$$\sum_{i=0}^{\infty} \Pi_i z^{-i} = D_u + H\left[\sum_{i=1}^{\infty} F^{i-1} z^{-i}\right] G_u$$

$$\sum_{i=0}^{\infty} \Lambda_i z^{-i} = D_e + H\left[\sum_{i=1}^{\infty} F^{i-1} z^{-i}\right] G_e \qquad \text{(B.8)}$$

$$\sum_{i=0}^{\infty} \Gamma_i z^{-i} = D_w + H\left[\sum_{i=1}^{\infty} F^{i-1} z^{-i}\right] G_w$$

where

$$\Pi_0 = D_u,\ \Pi_1 = HG_u, \ldots \Pi_k = HF^{k-1} G_u \ldots$$

$$\Lambda_0 = D_e,\ \Lambda_1 = HG_e, \ldots \quad \Lambda_k = HF^{k-1} G_e \ldots$$

$$\Gamma_0 = D_w,\ \Gamma_1 = HG_w, \ldots \Gamma_k = HF^{k-1} G_w \ldots$$

The final form matrices Π_i, Λ_i, Γ_i expressed in (B.8) are termed Markov matrices in the control literature. In the econometric literature, these matrices are matrix multipliers. Thus $\Pi_0 = D_u$, $\Lambda_0 = D_e$, are the impact multipliers of the controls and exogenous variables respectively, and measure the immediate effect of a unit variation in any input on any output.

Similarly, the matrices $\Pi_i\ (= HF^{i-1}G_u)$, $\Lambda_i\ (= GF^{i-1}G_e)$, $\Gamma_i\ (= HF^{i-1}G_w)$, $i = 1, 2, \ldots$, are called the dynamic matrix multipliers at lag i.

If step responses are considered, the partial sums

$$\Pi^j = \sum_{i=0}^{i=j} \Pi_i = D_u + \sum_{i=1}^{i=j} HF^{i-1} G_u$$

$$\Lambda^j = \sum_{i=0}^{i=j} \Omega_i = D_e + \sum_{i=1}^{i=j} HF^{i-1} G_e$$

$$\Gamma^j = \sum_{i=0}^{i=j} \Gamma_i = D_w + \sum_{i=1}^{i=j} HF^{i-1} G_w$$

$j = 1, 2, \ldots$, are cumulated matrix multipliers.

Since F is stable by assumption, the steady-state static matrix multipliers, representing the equilibrium effects on the outputs of a sustained unit change in the inputs are given by

$$\Pi^{\infty} = D_u + H[I - F]^{-1}G_u$$

$$\Lambda^{\infty} = D_e + H[I - F]^{-1}G_e.$$

These step responses for selected inputs and outputs are reproduced in Table B.1.

9 Adaptive Control of Macroeconomic Models with Measurement Error

David Kendrick[1]

The data used by policy analysts to determine monetary and fiscal policies are known to contain errors. In fact these data are constantly being revised as more information becomes available. The magnitude of these revisions gives us a measure of the relative quality of different macroeconomic time series and yet economists do not presently use this information in determining policies.

This paper reports on the beginnings of an investigation directed to the following questions about measurement errors:

1. Are the measurement errors indicated by these revisions of sufficient magnitudes to make any substantial difference in monetary and fiscal policies?

2. Will procedures designed to take account of the magnitude of these errors yield policies which are different from methods which ignore them?

3. When considering these measurement errors, will adaptive control procedures, which take account of the fact that parameters can be learned, result in substantially better policies than certainty equivalence control procedures, which ignore learning?

Rough calculations indicate that the answer to the first question is that measurement errors are large enough to make a difference. The answer to the second question has not yet been obtained but the procedures for answering it are now programmed and are almost ready to be used. Results obtained so far on the third question are mixed but presently indicate that the computationally expensive adaptive control procedures are somewhat better than the computationally simple certainty equivalence methods.

A description of procedures used and results obtained follows. This begins with a discussion of the model and data employed and is followed by a discussion of the method to be used in answering question 2 above. Finally, the results obtained so far on certainty equivalence versus adaptive control are given.

I. THE MODEL AND DATA

A simple multiplier-accelerator model in unobservable levels for desired consumption and investment can be written

$$C_k^* = a + bY_k \tag{1}$$
$$C_k = C_{k-1} + \gamma(C_k^* - C_{k-1}) \tag{2}$$
$$I_k^* = e + f(\Delta Y_k) \tag{3}$$
$$I_k = I_{k-1} + \theta(I_k^* - I_{k-1}) \tag{4}$$
$$Y_k = C_k + I_k + G_k \tag{5}$$

where

C_k = consumption
C_k^* = desired consumption
I_k = investment
I_k^* = desired investment
Y_k = GNP (net of net exports)
G_k = government spending.

The model (1)–(5) can be reduced to the following model in observable variables:

$$C_k = \alpha_0 + \alpha_1 Y_k + \alpha_2 C_{k-1} \tag{6}$$
$$I_k = \beta_0 + \beta_1 Y_k + \beta_2 Y_{k-1} + \beta_3 I_{k-1} \tag{7}$$
$$Y_k = C_k + I_k + G_k. \tag{8}$$

The reduced form of this model (estimated by OLS with data from the NBER data bank from 47II through 69I) is[2]

$$\begin{aligned} C_k = \underset{(.169)}{.852}C_{k-1} - \underset{(.140)}{.125}I_{k-1} + \underset{(.161)}{.156}Y_{k-1} - \underset{(1.53)}{1.482} - \underset{(.162)}{.158}\,G_k \qquad & R^2 = .99 \\ & DW = 2.22 \end{aligned} \tag{9}$$

$$\begin{aligned} I_k = -\underset{(.237)}{.328}C_{k-1} + \underset{(.196)}{.425}I_{k-1} + \underset{(.226)}{.403}Y_{k-1} + \underset{(.215)}{.008} - \underset{(.228)}{.499}\,G_k \qquad & R^2 = .94 \\ & DW = 1.66 \end{aligned} \tag{10}$$

$$Y_k = .527C_{k-1} + .301I_{k-1} + .557Y_{k-1} - 1.512 + .345G_k \qquad R^2 = .99$$
$$\quad (.293) \qquad (.243) \qquad (.280) \qquad (2.66) \quad (.282) \qquad DW = 1.12 \tag{11}$$

This model can be rewritten in the usual state space form as

$$x_{k+1} = Ax_k + Bu_k \quad k = 0, 1, \ldots, N-1 \tag{12}$$

where

$$x_k = \begin{bmatrix} C_k \\ I_k \\ Y_k \\ 1 \end{bmatrix} \quad u_k = G_{k+1} = L_k = \text{obligations,}$$

and 1 is simply the scalar one. The model assumes that government obligations can be controlled and that they lead government expenditures by one quarter. Also for this model

$$A = \begin{bmatrix} .852 & -.125 & .156 & -1.482 \\ -.328 & .425 & .403 & .008 \\ .527 & .301 & .557 & -1.512 \\ .0 & .0 & .0 & 1.0 \end{bmatrix}$$

$$B = \begin{bmatrix} -.158 \\ -.499 \\ .345 \\ .0 \end{bmatrix} \text{ and } x_0 = \begin{bmatrix} 460.1 \\ 113.1 \\ 718.4 \\ 1.0 \end{bmatrix}$$

The values used for x_0 are the actual values for the corresponding Series for 1969I as taken from the NBER Times Series Data Bank.

The criterion is written to minimise

$$J = (x_N - \rho_N)'W_N(x_N - \rho_N) + \sum_{k=0}^{N-1} [(x_k - \rho_k)'W_k(x_k - \rho_k) + (u_k - u_k^{\#})'\Lambda_k(u_k - u_k^{\#})] \tag{13}$$

where

ρ = desired state vector
$u^{\#}$ = desired control vector
W = weights on state deviations
Λ = weights on control deviations.

The paths ρ (except the last component of ρ) and $u^{\#}$ were chosen by assuming desired growth rates of .75 per cent per quarter. ρ_4, the last component of ρ, was set equal to one for all periods. The basis for these desired paths are the actual data for 1969I, i.e.[3]

$$\rho_0 = \begin{bmatrix} 460.1 \\ 113.1 \\ 718.4 \\ 1.0 \end{bmatrix} \qquad u_0{}^{\#} = [153.644].$$

The weighing matrices were chosen so as to give greater weights to state deviations in the terminal year than in other years in order to model the phenomena that political leaders care more about the state of the economy in quarters just before elections than in other quarters. Therefore, these matrices were set as

$W_N = (100, 100, 100, 0)$ $W_k = \text{diag}\,(1, 1, 1, 0)$ $\Lambda_k = 1$ $k = 1, \ldots, T-1$[4]

The stochastic version of the model is obtained by (i) minimising the expected value of (13) subject to systems equations

$$x_{k+1} = Ax_k + Bu_k + \xi_k \qquad \begin{matrix} x_0 \sim N(\bar{x}_0, \Sigma^{xx}_{0|0}) \\ \zeta_k \sim N(0, Q) \end{matrix} \tag{14}$$

and measurement relations

$$Y_k = Hx_k + \zeta_k \qquad \zeta_k \sim N(O, R) \tag{15}$$

where

ξ = systems equations noises
ζ = measurement errors
H = measurement matrix
y = measurement vector.

It is assumed here that the initial x is known imperfectly and that its estimate is normally distributed with mean $\bar{x}_0$ and covariance $\Sigma^{xx}_{0|0}$. (The reason for this elaborate notation for the covariance will be explained momentarily.)

The systems equation noise ξ_k and the measurement noise are both assumed to be normally distributed and serially uncorrelated with means zero and covariances Q and R respectively. It is not true that the error terms are uncorrelated. However, that assumption has been used here for the sake of simplicity. Kent Wall, Yaakov Bar-Shalom and I have begun a model which includes serial correlation in the measurement errors and hope to be able to show in the future how one can use this correlation to improve control of the economy.

The diagonal elements of the covariance of the systems noise terms, Q, are the square of the standard errors of the reduced form equation errors. The diagonal elements of this matrix are

$$Q = \text{diag}\,[9.61,\ 18.92,\ 28.94,\ 0.0].$$

The measurement error covariance, R, was estimated from the revisions data by the procedure outlined in Appendix A. The resulting matrix is

$$R = \begin{bmatrix} 2.71 & 1.12 & 5.52 & 0.0 \\ 1.12 & 2.78 & 5.42 & 0.0 \\ 5.52 & 5.42 & 16.22 & 0.0 \\ 0.0 & 0.0 & 0.0 & .0001 \end{bmatrix}$$

Note that the variance of the measurement error for consumption is low relative to its value of x_0, (2.71 on 460.1 billion), while that of investment is relatively high (2.78 on 113.1 billion). The algorithm described here takes account of this fact and relies less heavily on the observed value of investment (y_2) than the observed value of consumption (y_1) in updating estimates of both states and parameters and therefore in determining the control to be used in subsequent periods.

In the results reported here a single case of parameter uncertainty has been considered. In this case all five of the elements in the first row of A and B were learned (corresponding to the five parameters in the reduced form of the consumption function) and all other coefficients were assumed to be known perfectly. To solve this problem a parameter vector θ consisting of the uncertain parameters was created and added to the initial state vector x to create a new state vector z. The state equations for the augmented model are

$$x_{k+1} = A(\theta_k)x_k + B(\theta_k)u_k + \xi_k \tag{16}$$

$$\theta_{k+1} = D\theta_k + \gamma_k \tag{17}$$

where D is assumed to be an identity matrix and γ is assumed to have both mean and covariance equal to zero (to model the case of constant but unknown parameters).

Part of the systems equations, namely (16), is now nonlinear in the new state vector

$$z_k = \left[\frac{x_k}{\theta_k}\right]. \tag{18}$$

Also, the covariance of the state vector at time k as estimated with data obtained through period k is now defined as

$$\Sigma_{k|k}^{zz} = \left[\begin{array}{c|c} \Sigma_{k|k}^{xx} & \Sigma_{k|k}^{x\theta} \\ \hline \Sigma_{k|k}^{\theta x} & \Sigma_{k|k}^{\theta\theta} \end{array}\right] \tag{19}$$

Using this notation the initial conditions for the augmented state equations (16) and (17) are

$$\bar{x}_0 = \begin{bmatrix} 460.1 \\ 113.1 \\ 718.4 \\ 1.0 \end{bmatrix} \qquad \bar{\theta}_0 = \begin{bmatrix} .852 \\ -.125 \\ .156 \\ -1.482 \\ -.158 \end{bmatrix} \tag{20}$$

$$\Sigma_{0|0}^{xx} = \begin{bmatrix} 2.71 & 1.12 & 5.52 & 0.0 \\ 1.12 & 2.78 & 5.42 & 0.0 \\ 5.52 & 5.42 & 16.22 & 0.0 \\ 0.0 & 0.0 & 0.0 & .0001 \end{bmatrix} \tag{21}$$

$$\Sigma_{0|0}^{\theta x} = (\Sigma_{0|0}^{x\theta})' = \begin{bmatrix} 0.0 & 0.0 & 0.0 & 0.0 \\ 0.0 & 0.0 & 0.0 & 0.0 \\ 0.0 & 0.0 & 0.0 & 0.0 \\ .011 & .014 & .036 & 0.0 \\ 0.0 & 0.0 & 0.001 & 0.0 \end{bmatrix} \tag{22}$$

$$\Sigma_{0|0}^{\theta\theta} = \begin{bmatrix} .0287 & .0216 & -.0272 & .0241 & .0265 \\ .0216 & .0196 & -.0213 & .0033 & .0212 \\ -.0272 & -.0213 & .0262 & -.0285 & -.0258 \\ .0241 & .0033 & -.0285 & 2.3493 & .0385 \\ .0265 & .0212 & -.0258 & .0385 & .0314 \end{bmatrix} \tag{23}$$

The prior mean of x is set to x_0 and the prior mean for θ is set to the estimated reduced form parameter estimates for the consumption function. The state covariance (21) is set equal to the measurement error covariance. The covariance (22) was estimated by applying a Kalman filter to the same data that was used for estimating the reduced form of the model. The priors for the Kalman filter were identity matrices for Σ^{xx}

and $\Sigma^{\theta\theta}$ and a zero matrix for $\Sigma^{\theta x}$.[5] The procedure employed is described in detail in Appendix L of Kang and Kendrick (1976). The covariance (23) was estimated with the TSP econometric package.[6]

II. MAGNITUDE OF EFFECT OF MEASUREMENT ERROR ON POLICY[7]

Are measurement errors large enough that they potentially have a substantial effect on policy variables? One can obtain a first crude answer to this question by solving the deterministic optimal control problem with criterion (13) and systems equations (12) and obtaining thereby the feedback rule

$$u_k = M_k x_k + g_k, \tag{24}$$

where the feedback matrix M_k and vector g_k are obtained as the solution to the quadratic linear control problem. Then, an error, Δx, is reflected in Δu as

$$\Delta u_k = M_k \Delta x_k. \tag{25}$$

Using errors of one standard deviation for C, I, and Y and looking, for example, at the control policy for period 6 one obtains

$$\Delta u_6 = (.031\ .109\ .145) \begin{bmatrix} 1.64 \\ 1.66 \\ 4.02 \end{bmatrix} = .814 \text{ billion dollars.} \tag{26}$$

So a measurement error of one standard deviation could result in a policy change of approximately one billion dollars relative to a government budget in 1969 of 150 billion 1958 dollars or of two billion relative to the present budget levels of greater than 300 billion current dollars. In tax terms this could make a difference then between a tax cut of say 10 billion and 8 billion when tax cuts of this magnitude are being debated.

III. OUTLINE OF A MORE SOPHISTICATED TEST

A more sophisticated test of the effect of measurement errors is to do Monte Carlo runs to compare two approaches: (i) a 're-estimation' method ignoring measurement errors and (ii) an 'adaptive control' method including consideration of measurement errors. The 're-estimation' method is similar to the open-loop optimal feedback methods used by control engineers and amounts to the use of deterministic control methods (which ignore future learning) with quarterly re-estimation of the econometric model while ignoring measurement error. The 'adaptive control' method with measurement error employs the procedure suggested in Tse, Bar-Shalom and Meier (1973) and Tse and Bar-Shalom (1973). With this approach the model is re-estimated with an extended Kalman filter each quarter while taking measurement errors into account. Then the re-estimated model is used to compute optimal adaptive policies while considering future learning possibilities.

The core of the Tse and Bar-Shalom adaptive control method is in a search for the optimal control policy u_k^* in each period which will minimise the dynamic programming 'expected-cost-to-go'. The second order approximation to this cost-to-go can be separated into deterministic, cautionary and probing components.[8] For the problem at hand these three components of the optimal cost-to-go at time $N-k$ are:[9]

$$J^*_{d,N-k} = \min_{u_k} \{J_{D,N-k} + J_{C,N-k} + J_{P,N-k}\}$$

Deterministic: (27)

$$\begin{aligned} J_{D,N-k} = {} & \tfrac{1}{2}(u_k - u_k{}^{\#})' \Lambda_k (u_k - u_k{}^{\#}) \\ & + \tfrac{1}{2}(x_{O,N} - \rho_N)' W_N (x_{O,N} - \rho_N) \\ & + \tfrac{1}{2} \sum_{j=k+1}^{N-1} [(x_{0,j} - \rho_j)' W_j (x_{0j} - \rho_j) + (u_{oj} - u_j{}^{\#})' \Lambda_j (u_{oj} - u_j{}^{\#})] \end{aligned}$$

Cautionary: (28)

$$\begin{aligned} J_{C,N-k} = {} & \tfrac{1}{2} tr \{K^{xx}_{k+1} \Sigma^{xx}_{k+1|k}\} + \{K^{\theta x}_{k+1} \Sigma^{x\theta}_{k+1|k}\} \\ & + \tfrac{1}{2} tr \{K^{\theta\theta}_{k+1} \Sigma^{\theta\theta}_{k+1|k}\} \\ & + \tfrac{1}{2} \sum_{j=k+1}^{N-1} [tr\{K^{xx}_{j+1} Q_j\} + tr\{K^{\theta\theta}_{j+1} G_j\}] \end{aligned}$$

Probing: (29)

$$\begin{aligned} J_{P,N-k} = \tfrac{1}{2} \sum_{j=k+1}^{N-1} \Big\{ & tr[(A' K^{xx}_{j+1} B) \mu_j (B' K^{xx}_{j+1} A) \Sigma^{xx}_{j|j}] \\ & + 2tr[B' K^{xx}_{j+1} A)' \mu_j \{B'(K^{xx}_{j+1} f^x_\theta + K^{x\theta}_{j+1} D) + \Sigma e_i p^x b^i_\theta\} \Sigma^{\theta x}_{j|j}] \\ & + tr[\{(D' K^{\theta x}_{j+1} + f^{x'}_\theta K^{xx}_{j+1}) B + (\Sigma e_i p^x b^i_\theta)'\} \mu_j \\ & \{B'(K^{xx}_{j+1} f^x_\theta + K^{x\theta}_{j+1} D) + \Sigma e_i p^x b^i_\theta\} \Sigma^{\theta\theta}_{j|j}] \Big\} \end{aligned}$$

The deterministic component includes only the non-random components of the cost-to-go. The cautionary term depends only on $\Sigma_{k+1|k'}$ i.e. on the covariance matrices for the states x and parameter θ at time $k+1$ given data through time k. These covariances cannot be affected by future measurements but are affected by the choice of current control. For example, it may be true that a more vigorous control choice at time k will drive up these covariance matrices. The probing component (29) is a function of the post-measurement covariance matrices for all future time periods, $\Sigma_{j|j}$, i.e. it does depend on future measurements. In some cases a more vigorous control at time k might be expected to decrease these covariance matrices by increasing the magnitude of current states and thereby narrowing the standard errors of future state and parameter estimates. (As is shown in the results section of this paper this explanation is too simple but is serves well as a first approximation.)

From this outline one might expect adaptive control policies to be more vigorous than 're-estimation' policies in the early time periods of the model while it was learning the parameters and then less vigorous in later time periods. Results of this type are discussed later in this paper.

The code to carry out the comparison of re-estimation and adaptive methods has been mostly programmed and debugged but has not yet been exploited because of the interest of this investigator in a prior question: namely the comparison of certainly equivalence and adaptive methods.

IV. ADAPTIVE VERSUS CERTAINTY EQUIVALENCE POLICIES

When measurement errors are considered, will adaptive control methods yield substantially better results than certainty equivalence methods? Posed another way, one can ask whether or not it is worth while to carry out the elaborate calculations which are required to consider the possibility of learning parameters and the gains which accrue from this learning. Evidence obtained so far has yielded seven Monte Carlo runs in which the adaptive method is superior and three in which the CE method is better.

The computer codes from which these results have been obtained are rather complex since they invoke both estimation and optimisation procedures embedded in a Monte Carlo routine, so independently coded programs are being used in our project to check results. Fred Norman, Yaakov Bar-Shalom and this investigator have independently coded various versions and parts of the adaptive algorithms. The most complicated part of the code is in the evaluation of the cost-to-go. Norman (using his program) and Kent Wall (using Bar-Shalom's programs) and I have been able to duplicate one another's results on a number of other problems, but we have not yet had the time and resources to fully check these problems. Therefore, the results presented here must be checked against one's intuition until such time as we can complete the numerical checking. It is in this spirit that the following results are presented.

Table 9.1 shows the results from the ten Monte Carlo runs completed. For each run random values of the systems noise ξ_k, the measurement noises ζ_k, the initial state estimate $\bar{x}_\theta$, and the initial parameter estimate $\bar{\theta}_\theta$ are generated using the means and covariances described above.

TABLE 9.1 Total Cost

Comparison of Dual and Certainty Equivalence		
Run #	*DUAL*	*CE*
1	36,720	12,690
2	68,060	47,830
3	20,400	52,660
4	36,400	6,304
5	29,790	59,010
6	15,140	16,570
7	31,790	61,920
8	148,000	207,500
9	22,120	30,510
10	34,300	40,040

The results of the first of the ten Monte Carlo runs are presented in Figures 9.1–9. For this particular run the CE solution was better than the DUAL run; none the less, it permits a demonstration of the type of results obtained. Figures 9.1–3 show the time paths for the state variables C, I, and Y. Figure 9.4 shows the results for the control variable G (or really obligations, L). Figures 9.5–9 show the time paths of the parameter estimates for the five parameters in θ, namely a_{11}, a_{12}, a_{13}, a_{14}, and b_1. For the state and controls, the figure also shows the desired paths ρ and $u^{\#}$ respectively and for the parameters the figures show the true value as well as the estimates obtained by the two methods.

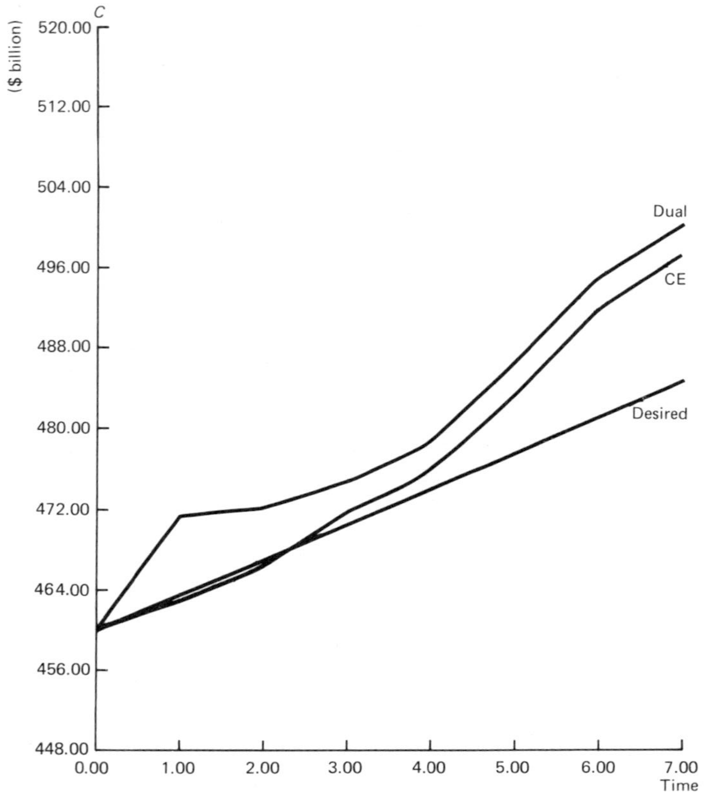

FIG. 9.1 Consumption

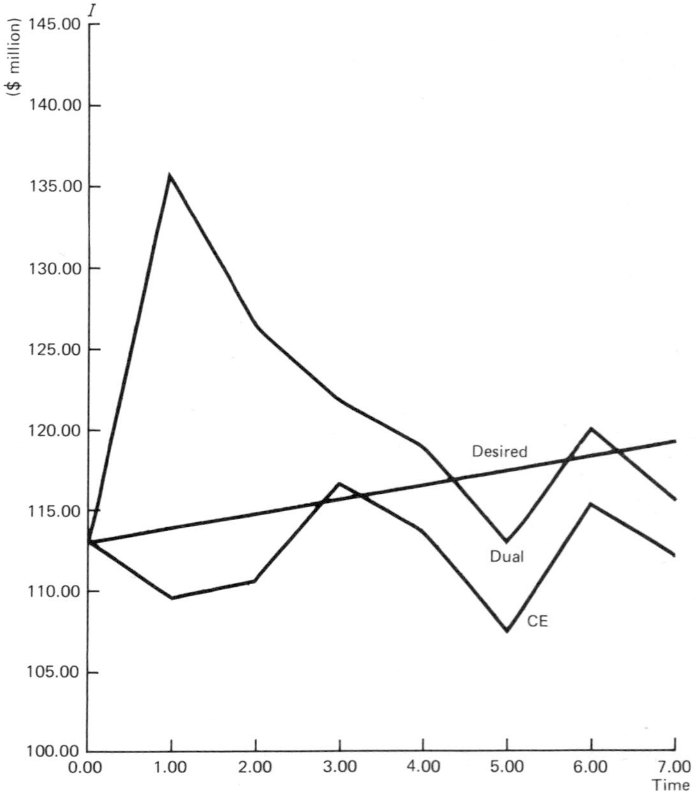

FIG. 9.2 Investment

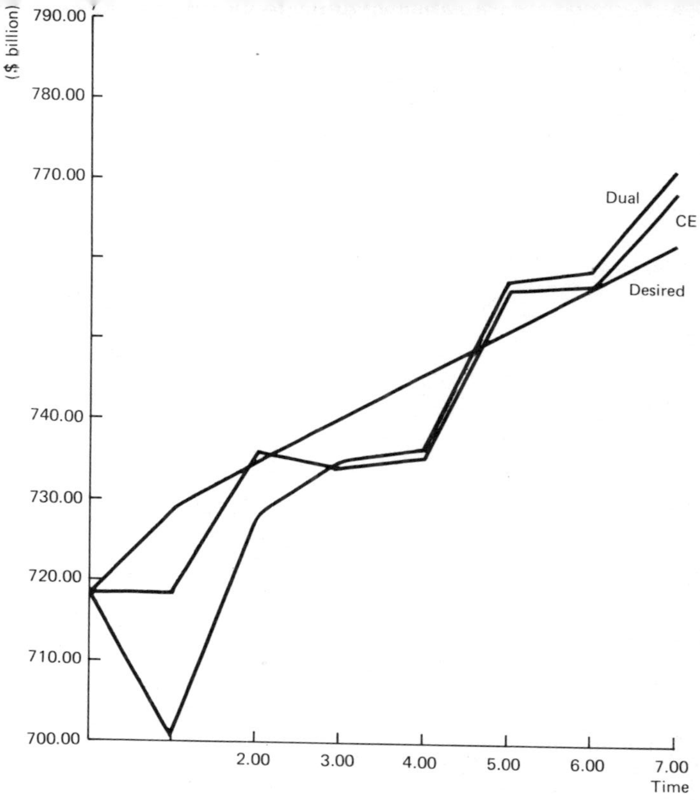

FIG. 9.3 GNP

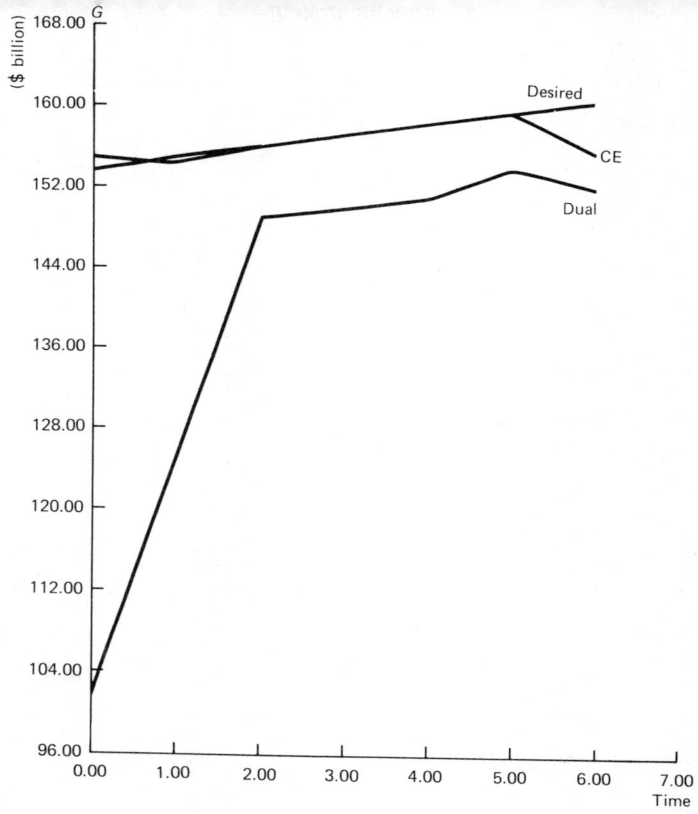

FIG. 9.4 Government Spending

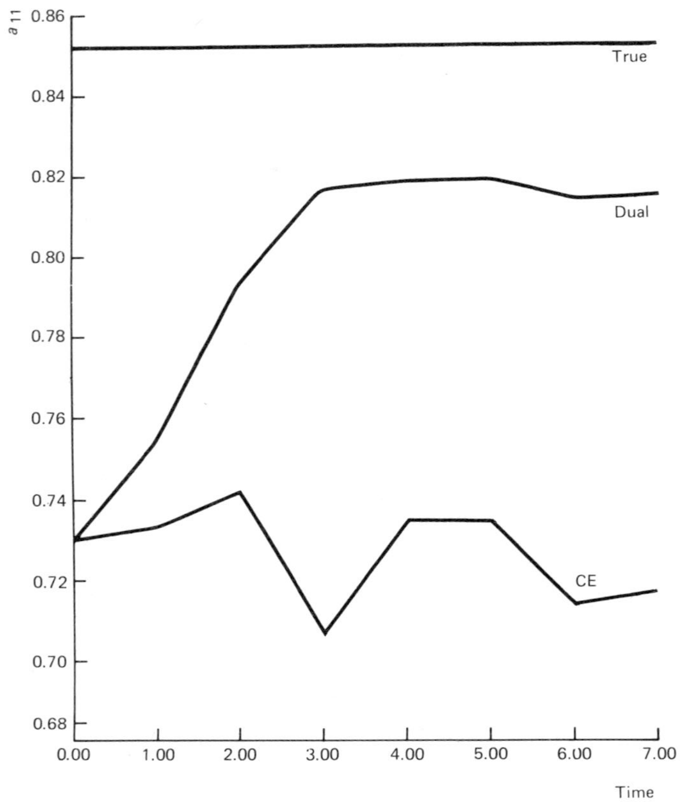

FIG. 9.5 Path of Parameter a_{11}

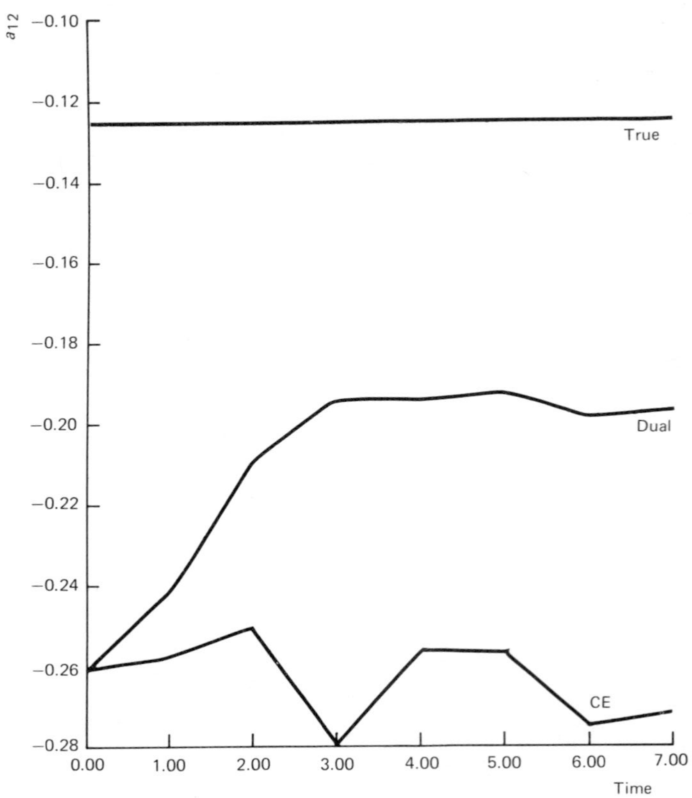

FIG. 9.6 Path of Parameter a_{12}

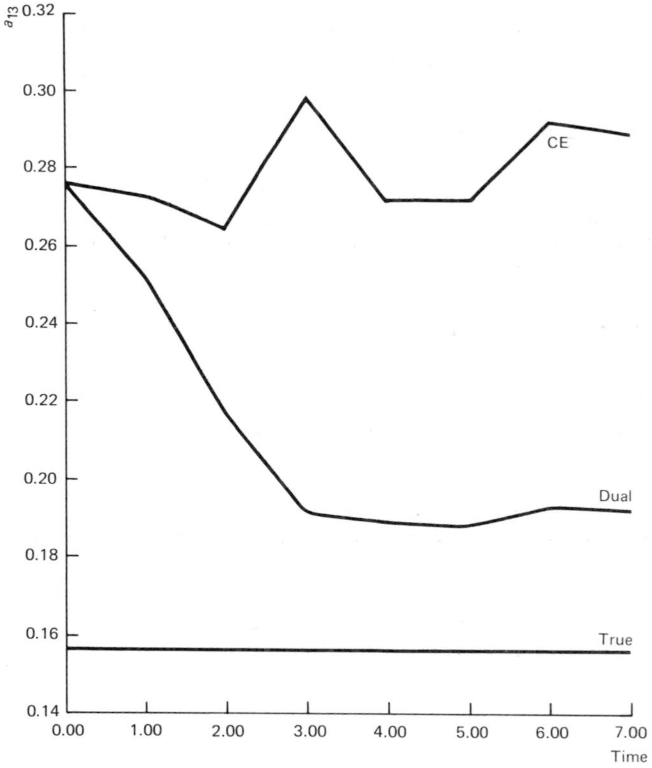

FIG. 9.7 Path of Parameter a_{13}

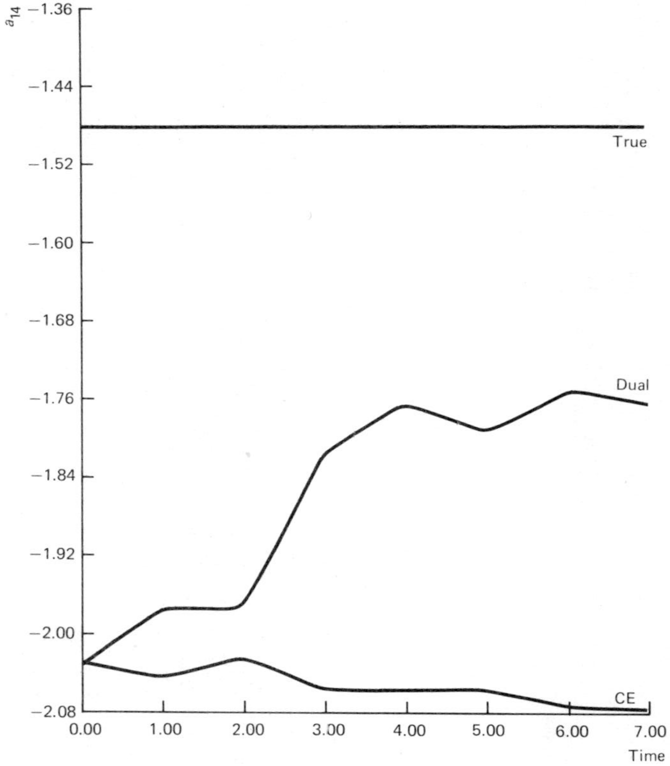

FIG. 9.8 Path of Parameter a_{14}

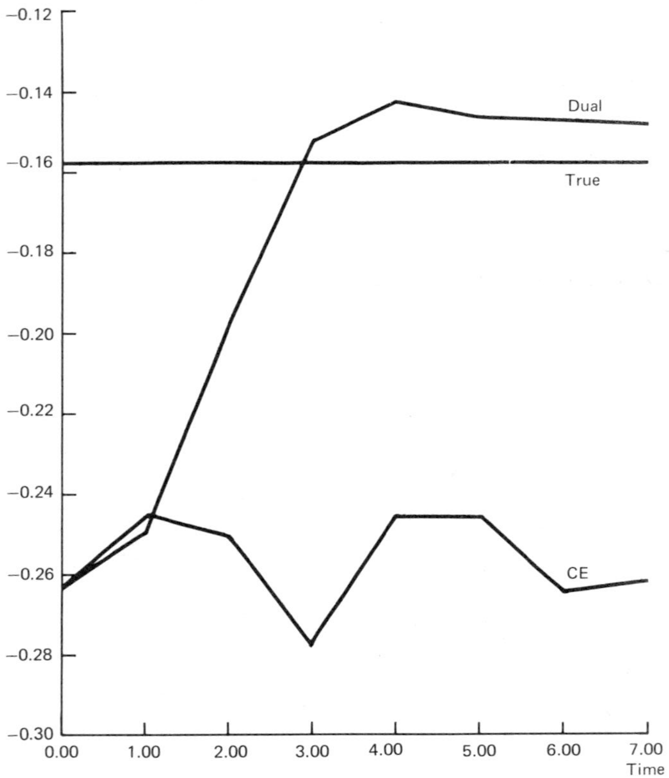

FIG. 9.9 Path of parameter b_1

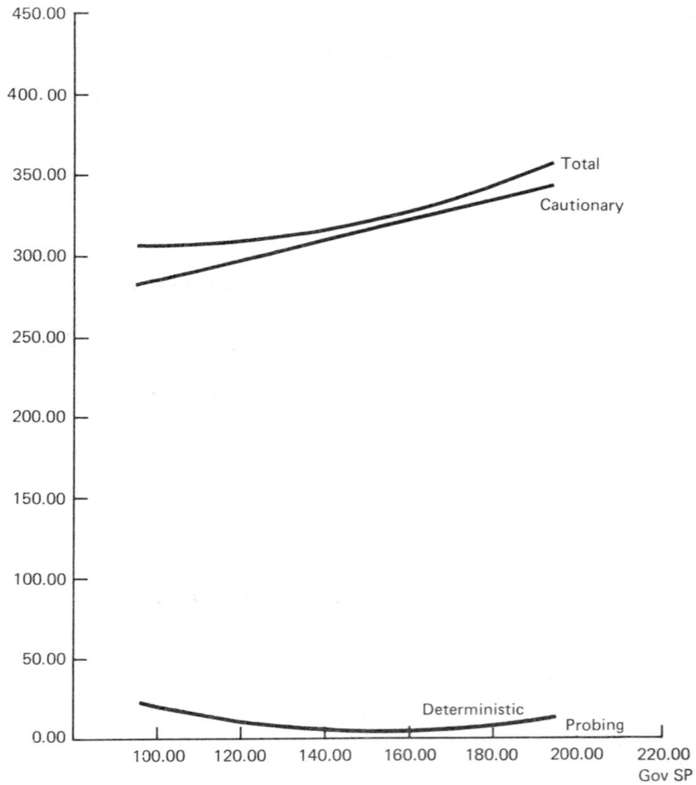

FIG. 9.10 Total Cost-to-go at Period 0

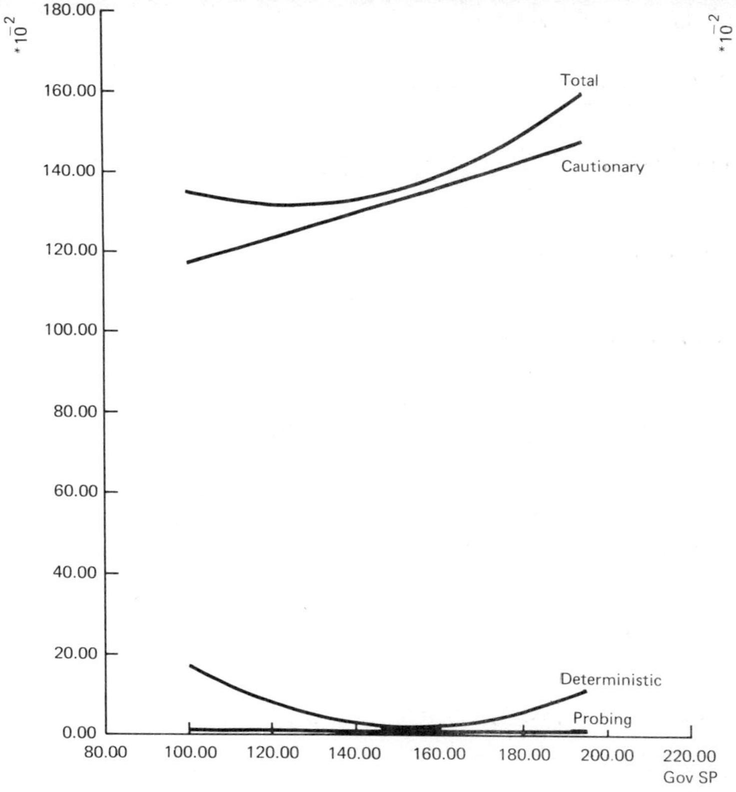

FIG. 9.11 Total Cost-to-go at Period 1

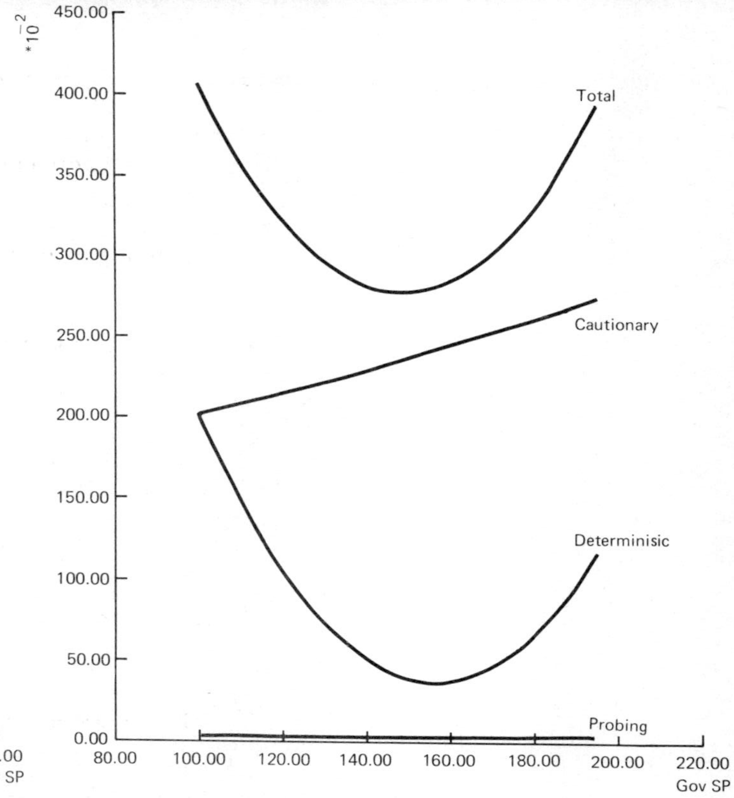

FIG. 9.12 Total Cost-to-go at Period 2

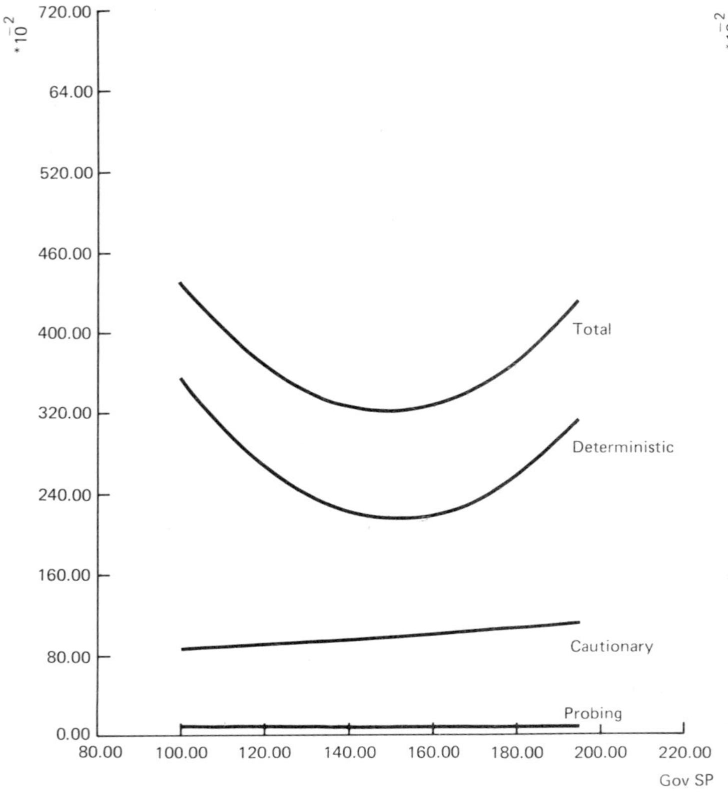

Fig. 9.13 Total Cost-to-go at Period 3

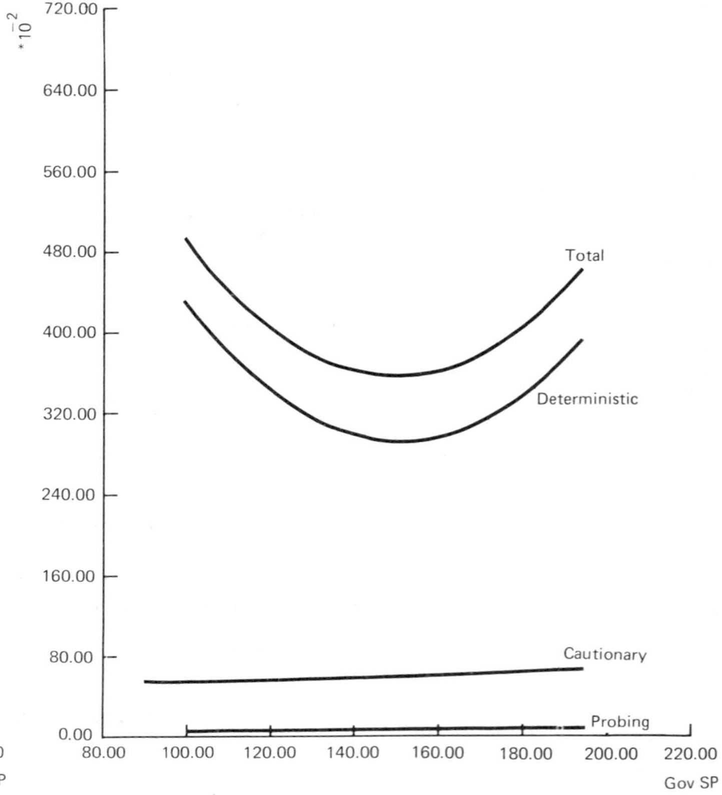

Fig. 9.14 Total Cost-to-go at Period 4

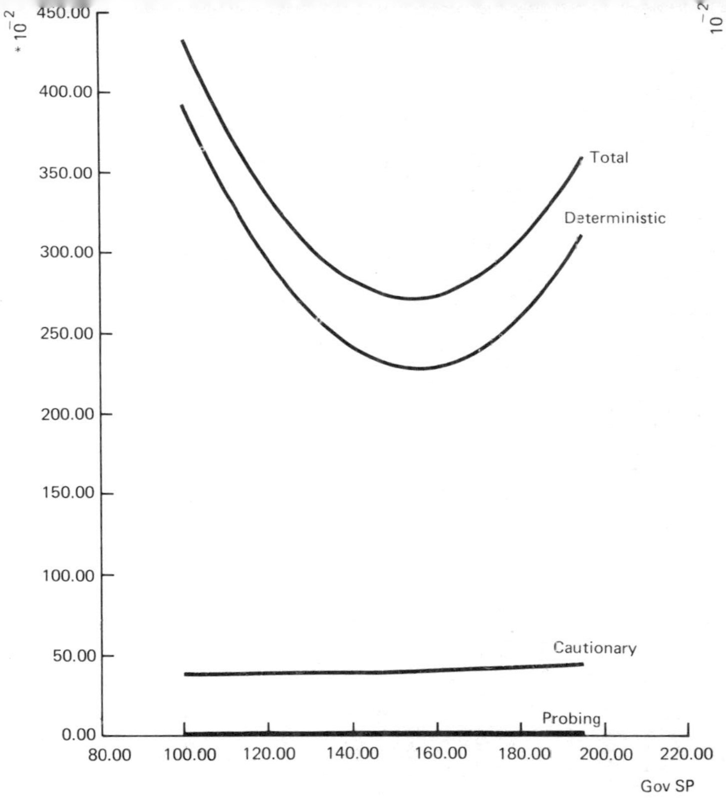

FIG. 9.15 Total cost-to-go at Period 5

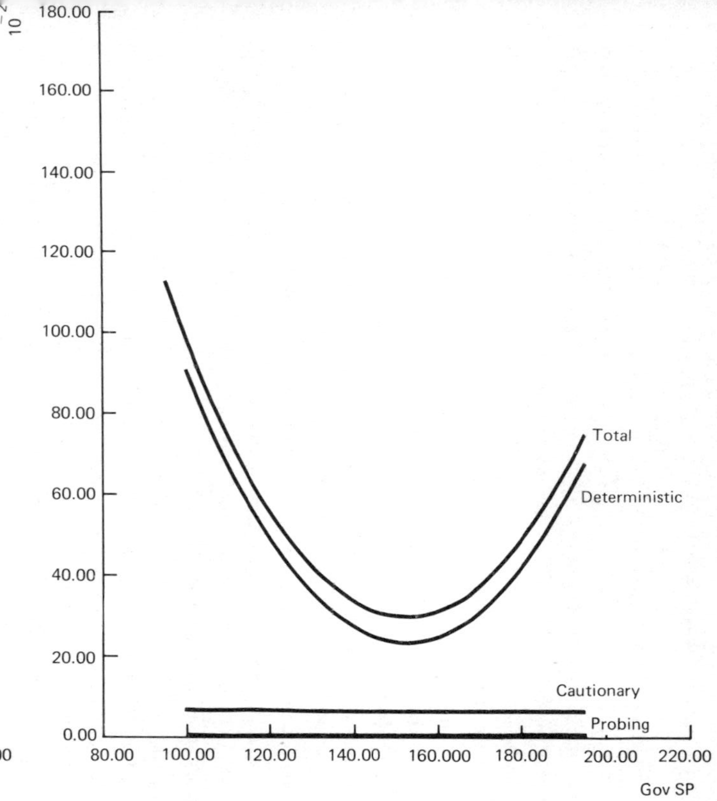

FIG. 9.16 Total Cost-to-go at Period 6

Figure 9.4 shows that the DUAL method chooses a very low level of government spending in the first two periods. As will be shown in a moment, this is primarily due to the cautionary cost term in our breakdown of the cost-to-go. Thus it appears that the DUAL methods choose a very cautious control in the first period when it knows relatively little about the system (i.e. the covariances in Σ are large). This has the effect of causing the DUAL method to rapidly improve the parameter estimates in the first three periods—see Figures 9.5–9.

Figure 9.4 also illustrates that in subsequent experiments the penalty on deviations from the desired control path should be increased relative to the penalty on deviations from the state path. The reason for this is that one would like to answer the question of whether or not mild perturbations away from the desired path for government spending will ever supply enough additional information to policy analysts to make such perturbations worth while.

Figures 9.10–16 show the total cost-to-go by period and its breakdown into deterministic, cautionary and probing terms as a function of the control u_k (government spending).[10] The cautionary component is the dominant element in the first few time periods. It is both large in magnitude and has a decided positive slope. This results in substantially smaller levels of the control in the first few periods than would result from the deterministic element alone. For example, the deterministic optimum level of government spending in period 0 appears to be about $150 billion while the minimum for the total cost-to-go is around $100 billion.

By way of contrast, the probing cost term is very small and has almost no slope, therefore it has little or no effect on the optimal level for the control. This is true throughout the seven time periods. The probing cost term is ever so slightly curved in a non-convex fashion but is never large nor does it ever have a pronounced slope.

As might be expected the cautionary term becomes of less and less importance as one progresses from period 0 to period 6 and it appears to have a smaller positive slope with each passing period.

These results have been obtained too recently to permit detailed analysis of the components of the cost-to-go for all ten Monte Carlo runs. However, it appears that the general results hold true, namely, that (i) the probing term is never of much importance, (ii) the cautionary term dominates in the early periods, and (iii) the deterministic term dominates in later periods.

A ratio of five or ten to one for the size of the terminal state deviation penalty relative to the other period state deviations seems reasonable (i.e.

ratio of elements in W_N to elements in W_k). That is, it seems likely that politicians care a lot more about the state of the economy in the quarters just prior to an election than they do in other periods. In these runs this ratio has been set at 100 rather than at 5 or 10 in order to see if deliberate probing would occur. Indeed, learning does occur but it appears that this learning is due to the effects of the cautionary term and not to the size and slope of the probing term.

V. CONCLUSIONS

Using statistical revision as an indication of the size of measurement errors in the US national income accounts and a deterministic feedback control model, it appears that the measurement errors are large enough to make significant differences in policy levels.

A more sophisticated attempt to answer this same question has been programmed but the results have not yet been obtained.

When employing methods which consider measurement errors, results obtained so far indicate that dual control results appear to hold an edge over certainty equivalence results.

NOTES

1. This research was supported by National Science Foundation grants SOC 72–05254 and SOC 76–11187. I would like to thank Yaakov Bar-Shalom, Fred Norman, Edison Tse, Kent Wall, Jorge Rizo-Patron and Steve Larson for comments and suggestions and Peggy Mills for administrative and typing help.
2. The actual NBER variables used were

C_k = GC58:	Personal Consumption Expenditures, billions of 1958 dollars,
I_k = GPI58:	Gross Private Domestic Investment, billions of 1958 dollars,
Y_k = GNP58– GNET 58:	Gross National Product – Net Exports of Goods and Services, billions of 1958 dollars,
G_k = GGE58:	Total Government Purchases of Goods and Sevices, billions of 1958 dollars.

3. A mistake was made in calculating the desired path for ρ_3. The initial ρ_3 is 718.4 as reported here but ρ_3 in period one is 728.926. The ρs for subsequent periods then grow at .75 per cent per quarter from this basis in period one.
4. The 100 to 1 ratio is too large but has been used here so far to test the algorithm and obtain first results.

5. See 9 Nov 76 0620 run.
6. The $\Sigma^{\theta\theta}$ obtained from TSP was not positive definite (probably due to roundoff errors in TSP) so .05 was added to the (5.5) element to increase it from .0264 to .0314.
7. I am indebted to Yaakov Bar-Shalom for suggesting this procedure.
8. See Bar-Shalom and Tse (1976) and Tse and Bar-Shalom (1976).
9. See Kang and Kendrick (1976), pp. 8–9.
10. A grid search method was used to obtain the points shown in these figures. First the functions were evaluated at 20 points between $u_{k'} = 100$ and $u_k = 195$. Then the function was evaluated at ten points around the minimum found in the first grid search.

REFERENCES

1. Bar-Shalom, Yaakov, and Edison Tse (1976), 'Caution, Probing and the Value of Information in the Control of Uncertain Systems', *Annals of Economic and Social Measurement*, vol. 5:2 (spring).
2. Conrad, William E. (1977), 'Imperfect Observation and Systematic Policy Error', *Annals of Economic and Social Measurement*, vol. 6:2.
3. Kendrick, David, and Bo Hyun Kang (1975), 'A Guide to Wide Sense Dual Control', Paper No. 75–1, Project on Control in Economics Dept. of Economics, University of Texas.
4. Kang, B. H., and D. Kendrick (1976), 'A Guide to Adaptive Control of Linear Systems with Random Parameters', Paper No. 75–4, revised Oct 1976, Project on Control in Economics, Dept. of Economics, University of Texas.
5. Tse, E., and Y. Bar-Shalom (1973), 'An Actively Adaptive Control for Linear Systems with Random Parameters', *IEEE Transactions on Automatic Control*, AC–18, pp. 109–17.
6. Tse, E., and Y. Bar-Shalom (1976), 'Adaptive Control for Nonlinear Stochastic Systems', *Proceedings of the IEEE*, vol. 64, no. 8, pp. 1172–81 (Aug).
7. Tse, E., Y. Bar-Shalom and L. Meier (1973), 'An Actively Adaptive Control for Linear Systems with Random Parameters', *IEEE Transactions on Automatic Control*, AC–18, pp. 98–108 (Apr).

Appendix A The Measurement Error Covariance

Following the work of Conrad (1977), the revisions of the national income accounts were used to obtain an estimate of the covariance matrix of the noise term of the measurement equations. This was done by assuming that the latest revision available was the true value and that the difference between this and the initial estimate is the size of the measurement error.

Table 9A.1 below gives the first reported value of GC58, GPI58, and GNP58, and Table 9A.2 gives the latest revision used in this study (those published in the *Survey of Current Business* on or before the November 1968 issue).

The difference between these two series, of course, understates the magnitude of the true measurement error. Worse still they may provide misleading estimates of the true measurement error since those series

TABLE 9A.1 First Reported Value (billions of 1958 dollars)

Quarter	*GC58*	*GPI58*	*GNP58*
64I	364.5	83.8	567.1
64II	369.8	85.2	575.9
64III	377.3	86.0	582.6
64IV	376.8	90.2	584.7
65I	385.9	94.7	597.5
65II	390.2	93.0	601.4
65III	396.7	92.9	609.7
65IV	403.3	100.5	624.4
66I	409.9	100.9	633.6
66II	412.2	106.3	643.5
66III	418.3	102.5	649.5
66IV	418.5	106.4	657.2
67I	422.0	95.7	656.7
67II	430.6	91.3	664.7
67III	431.5	96.4	672.0
67IV	434.0	103.0	679.6

that have the largest true measurement error may be the most difficult to revise and thus be the series that shows the least revision and therefore the smallest error. So, the measurement errors shown in the revisions in Table 9A.3 reflect lower bounds on the true measurement error. As this

TABLE 9A.2 Latest Revision (billions of 1958 dollars)

Quarter	*GC58*	*GPI58*	*GNP58*
64I	366.3	85.3	571.1
64II	370.7	87.3	578.6
64III	378.6	87.6	585.8
64IV	379.3	90.8	588.5
65I	387.9	96.9	601.6
65II	393.4	96.8	610.4
65III	400.3	99.6	622.5
65IV	409.2	103.4	636.6
66I	415.7	106.1	648.6
66II	414.8	109.5	653.3
66III	420.0	107.4	659.5
66IV	420.6	112.3	667.1
67I	424.8	99.8	665.7
67II	431.2	94.2	669.2
67III	431.8	99.3	675.6
67IV	434.1	104.7	681.8

TABLE 9A.3 Size of Revisions (billions of 1958 dollars)

Quarter	*GC58*	*GPI58*	*GNP58*
64I	−1.8	−1.5	−4.0
64II	−0.9	−2.1	−2.7
64III	−1.3	−1.6	−3.2
64IV	−2.5	−0.6	−3.8
65I	−2.0	−2.2	−4.1
65II	−3.2	−3.8	−9.0
65III	−3.6	−6.7	−12.8
65IV	−5.9	−2.9	−12.2
66I	−5.8	−5.2	−15.0
66II	−2.6	−3.2	−9.8
66III	−1.7	−4.9	−10.2
66IV	−2.1	−5.9	−9.9
67I	−2.8	−4.1	−9.0
67II	−0.6	−2.9	−4.5
67III	−0.3	−2.9	−3.6
67IV	−0.1	−1.7	−2.2

kind of work proceeds it will be useful to attempt to obtain independent information on the magnitudes of the measurement errors by doing in depth studies on some elements of the time series.

A glance at Table 9A.3 confirms that the revisions are serially correlated and have non-zero means. However, for purposes of this study, we have assumed that the measurement error has zero means and is uncorrelated over time. A later study will attempt to exploit the information in the serial correlation and non-zero means of these statistics.

The covariance of these time series is given in Table 9A.4. With the addition of a bordering row of zeros to reflect the fact that there is no error in measuring the constant term, 1, this is the 4 × 4 matrix used for *R*, the covariance of the measurement noise.

TABLE 9A.4 Covariance of Revisions

	GC58	*GPI58*	*GNP58*
GC58	2.71	1.12	5.52
GPI58	1.12	2.78	5.42
GNP58	5.52	5.42	16.22

There is a slight error in the components for GNP58 since the model actually used are GNP58-GNET58, i.e., GNP net of net exports. However, the magnitude of this difference is small.

PART IV

STABILISATION POLICIES WITH LARGE NONLINEAR ECONOMETRIC MODELS

10 Control of the LBS Econometric Model via a Control Model

M. B. Zarrop, S. Holly, B. Rüstem,
J. H. Westcott and M. O'Connell

1. INTRODUCTION

As a result of the work that has been devoted to the use of control theory in macroeconomics, it has become clear that demonstrating the basic principles on small models, though useful as a pedagogic exercise, is not likely by itself to achieve the needed transition from theory to actual practice in economic policy-making. When a control engineer has to convince industrialists that he can make a contribution to the solution of their problems he is able to do so, in the final analysis, by showing that the performance of the boiler or the chemical process has been improved. This can be achieved in a relatively short space of time. In the management of the economy, however, the period necessary to demonstrate that stochastic control methods are beneficial may be prohibitively long. There is, therefore, a serious problem of validation since the Government are not going to hand the economy over so that experiments in control can be conducted. One way this obstacle can be avoided is to use a large nonlinear econometric model of the economy, such as that of the London Business School, as an analogue for the real economy, and to control it with a small linear model estimated from economic data. Needless to say, the large model itself is not free from many of the criticisms that can be levelled at the small model. But as long as the model, in terms of its dynamic responses, behaves like the economy, it will provide an adequate test-bed for the use of control methods.

For a prescribed objective function, the basic idea is to control the larger model by applying a feedback law that is optimal for some simple model. It is shown that the structure of the PREM model can be exploited to yield a simple scheme for implementing this idea. Preliminary results of the 'link-up' using a quadratic welfare function to penalise growth rates in certain economic variables, have been encouraging. In particular, the results indicate the inadequacies of applying policy in an open-loop (simulation) mode rather than via a closed-loop feedback law.

2. REDUCED-ORDER MODELS

The vast majority of attempts at controlling systems via reduced-order models proceed from some given linear state-space representation of the system. Aoki (1971) develops a state aggregation scheme to produce an aggregate model tailored to the control exercise to be performed. Other authors incorporate the dominant system eigenvalues into their models (Chidambara, 1969) and/or match the model final form equations to the system by an expansion in the differential (or delay) operator (Shamash, 1975) in order to capture the long-term behaviour of the system. Such schemes are not immediately applicable to the control of the large nonlinear models used for economic forecasting, but they indicate the type of 'structural condensation' that is necessary if the policy objectives are precisely defined and are to be realised by employing a control model of relatively low dimension. An important point is that, in general, the small model that emerges is dependent on the policy objectives. Perhaps most criticism of such schemes will come from economists who will plead that the economy can in no way be viewed as a 'given system' and that they have a very limited degree of confidence in any macroeconomic forecasting model. One way of avoiding large model dependence is to estimate the simple model directly from economic data while ensuring that all important instruments and targets are present. This is the approach developed here and it may be argued that it is a reasonable one to investigate given that policy-makers do not have some 'true' model of the economy which can then be condensed to their requirements but must rely on given economic data.

3. THE PREM MODEL STRUCTURE

The PREM models are linear, stochastic and time-invariant, estimated in rational structural form from variables expressed in percentages changes. In state space form (Preston and Wall, 1973), the model can be written as

$$\underline{x}(k+1) = F\underline{x}(k) + G_u\underline{u}(k) + G_e\underline{e}(k) + G_w\underline{w}(k) \tag{3.1a}$$
$$\underline{y}(k) = H\underline{x}(k) + D_u\underline{u}(k) + D_e\underline{e}(k) + D_w\underline{w}(k) \tag{3.1b}$$

where the linearised identities have been substituted out and, at time k,

$\underline{x}(k)$ = state vector (50), $\underline{y}(k)$ = output vector (20)
$\underline{u}(k)$ = vector of instruments (8)
$\underline{e}(k)$ = vector of exogenous variables (7)
$\underline{w}(k)$ = vector of Gaussian stochastic disturbances (20)

The appended numbers denote the dimension of each vector for the PREM model (Holly *et al.*, Chapter 8 of this volume).

The special structure of (3.1) has implications for state estimation and optimal control.
Eliminating $\underline{w}(k)$ from (3.1) yields

$$\underline{x}(k+1) = A\underline{x}(k) + B\underline{u}(k) + C\underline{e}(k) + K\underline{y}(k) \tag{3.2}$$

where

$$K \triangleq G_w D_w^{-1}, \quad A \triangleq F - KH$$
$$B \triangleq G_u - KD_u, \quad C \triangleq G_e - KD_e$$

Thus, if the input-output data is given and $\underline{x}(k)$ is known for $k = k_o$, say, then (3.2) allows the exact calculation of the state for all $k > k_o$. If $\underline{x}(k_o)$ is in error, then the error will decay away if A is stable, i. e. has all its eigenvalues inside the unit circle. This condition is equivalent to the stability of the mapping $y \rightarrow w$ (Preston and Wall, 1973) and this is ensured by the estimation procedure (Wall, 1976). Hence it can be argued (Wall and Westcott, 1975) that if $\underline{x}(-d)$ is (erroneously) assumed to be zero for some $d > o$ (preferably at the beginning of the historical data record) then using (3.2) to roll forwards over historical data will yield an accurate value for $\underline{x}(1)$, the state at the control origin $k = 1$. Once $\underline{x}(1)$ is assumed known, then the state can be calculated exactly throughout the control period, thus avoiding laborious state estimation techniques (e.g. Kalman filtering).

The validity of the assumption of full state information depends on the size of $|\lambda_{\max}|^d$, where $\lambda_{\max}$ is an eigenvalue of A with maximum modulus. For the latest PREM model,

$$\lambda_{\max} = 0.734 \pm 0.597j \quad |\lambda_{\max}| = 0.946$$

and, taking the control origin at 70(1), only 22 periods of full historical data are available so that

$$|\lambda_{\max}|^d > (0.946)^{22} = 0.295$$

In this case, the full state information assumption is on rather a weak footing and it would be preferable to use Kalman filtering techniques. The simplicity of (3.2), however, and its implications for the link-up algorithm below make it worth while accepting $\underline{x}(1)$ as the exact initial state, at least for these preliminary experiments.

The solution of the optimal control problem for PREM models when the state is known and a quadratic objective function is used has been reported elsewhere (Bray, 1975; Wall and Westcott, 1975). The suite of programs for producing the state space realisation (3.1) and deriving the optimal feedback law has been substantially revised to accommodate larger models and to allow for time-varying weights. The optimal policy, i.e. that which minimises the objective function,

$$J = E\{\tfrac{1}{2} \sum_{k=1}^{N} [\delta\underline{y}(k)^T A(k)\delta\underline{y}(k) + 2\delta\underline{y}(k)^T B(k)\delta u(k) + \delta\underline{u}(k)^T C(k)\delta\underline{u}(k)]\} \tag{3.4}$$

(where $E\{.\}$ denotes the expectation operator and δ denotes deviation from the desired value) is expressible in the form

$$\underline{u}(k) = L(k)\underline{x}(k) + \underline{h}(k),\ k,\ 1, \ldots, N \tag{3.3}$$

The arrays $L(k)$, $\underline{h}(k)$ are termed the control and tracking gains respectively and N is the control horizon. Only $\underline{h}(k)$ is dependent on the desired trajectories and exogenous assumptions.

4. CONTROL OF THE LBS MODEL

The control of of the LBS model is depicted in Figure 10.1. The symbols $\underline{U}(k)$, $\underline{Y}(k)$ are used to denote respectively the vectors of instruments and endogenous variables in levels for the LBS model at time k.

The model to be controlled is nonlinear and deterministic. It is estimated in levels and expressed in polynomial structural form. Aggregation and differencing operations and their inverses are necessary to ensure compatibility between $\{\underline{U}(k), \underline{Y}(k), k = 1, \ldots, N\}$ and $\{\underline{u}(k), \underline{y}(k), k = 1, \ldots, N\}$. These operations are written symbolically as

$$\underline{y}(k) = L_y\underline{Y}(k);\ \underline{U}(k) = L_u\underline{u}(k) \tag{4.1}$$

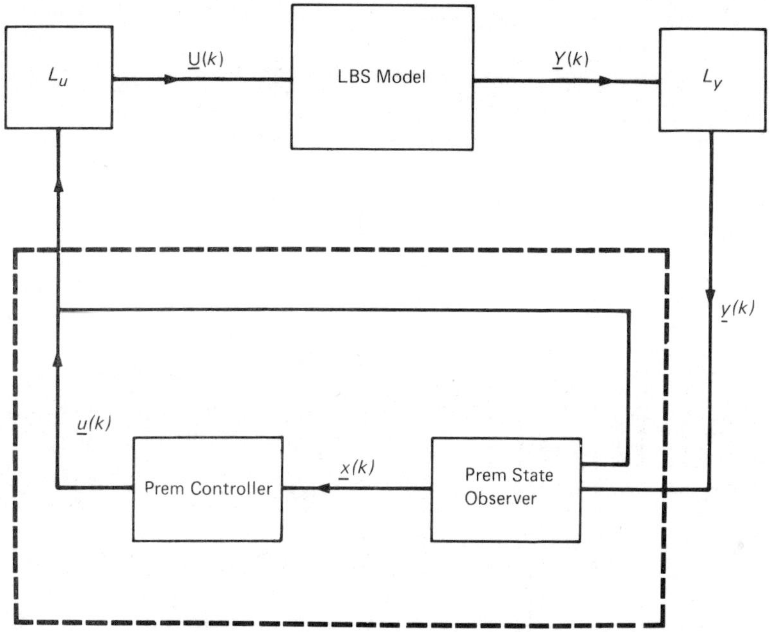

FIG. 10.1 Diagram of LBS/PREM Control Link-up

The analysis of the last section leads to the detailed expansion of Figure 10.1 into 10.2 and the following computational algorithm:

Phase 1 For the PREM model:
(*a*) Set desired targets for control and output variables and exogenous assumptions $\underline{e}(k)$, $k = 1, \ldots, N$;
(*b*) Set weights for the quadratic cost function according to priorities;
(*c*) Run the PREM stochastic control program to obtain the optimal policy;
(*d*) Store the control and tracking gains and the initial state $\underline{x}(1)$.

Phase 2 For the LBS model;
(*a*) Set $i = o$
(*b*) Set $i = i + 1$
(*c*) Set $\underline{u}^*(i) = L(i)\underline{x}(i) + \underline{h}(i)$ and input $\underline{U}(i) = L_u \underline{u}^*(i)$ to the LBS model to generate $\underline{Y}(i)$ and hence $\underline{y}(i) = L_y \underline{Y}(i)$.
(*d*) If $i = N$, STOP.
(*e*) Set $\underline{x}(i+1) = A\underline{x}(i) + B\underline{u}^*(i) + C\underline{e}(i) + K\underline{y}(i)$.
(*f*) Go to (*b*).

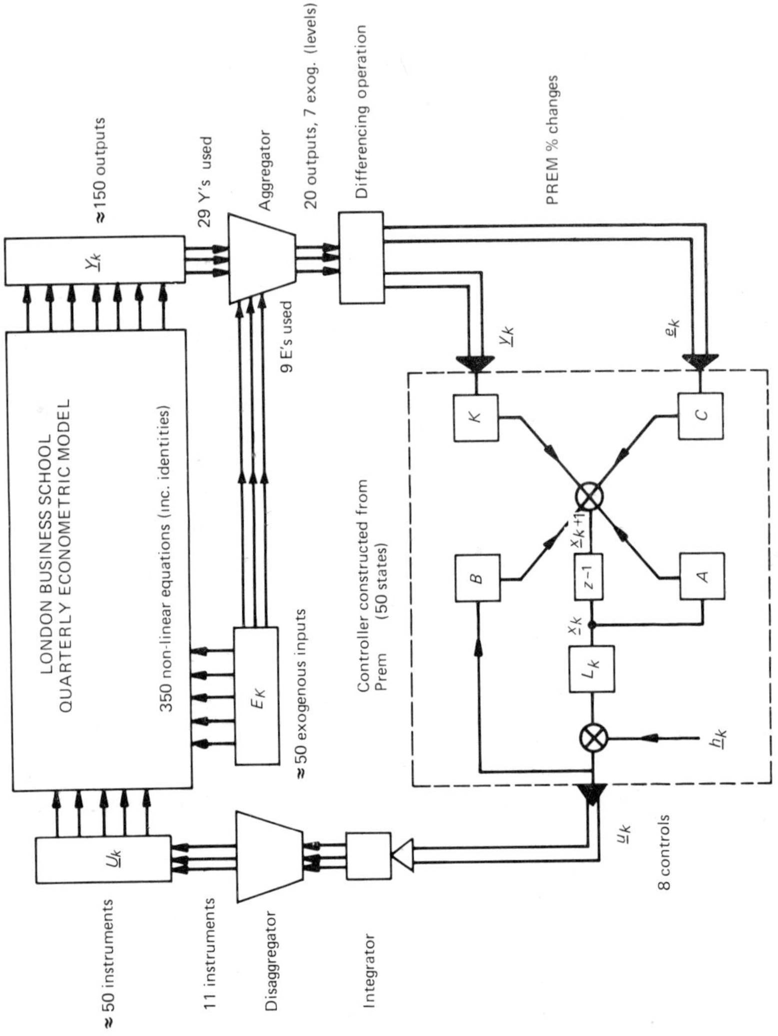

FIG. 10.2 Control of the LBS model using the PREM model

5. A NOTE ON INPUT DISAGGREGATION

During the development of a PREM model suitable for the purposes of the link-up, the problem of input disaggregation was raised. In its simplest form the problem is to choose $U(k)$, such that

$$\underline{u}(k) = A_o \underline{U}(k) \tag{5.1}$$

where A_o is some given 'aggregation matrix' and $\underline{u}(k)$ is the control produced by PREM and converted to levels for inputing to the large model.

In general, dim $[\underline{u}(k)] <$ dim $[\underline{U}(k)]$ and, therefore, the disaggregation of $\underline{u}(k)$ into $\underline{U}(k)$ does not yield a unique control but an infinite number of possibilities. The simplest path, and that chosen by PREM, is to identify the key components of $\underline{U}(k)$ directly with those in $\underline{u}(k)$ and to exogenise the others. It is possible, however, to take up the 'slackness' in $\underline{U}(k)$ by allowing the policy-maker to take account of some of his secondary requirements which may not be reflected in the aggregated criterion used to control the PREM model.

Assuming that the policy-makers' preferences concerning the levels of the instruments at time k are embodied in the objective function

$$J(k) = [\underline{U}(k) - \underline{U}_d(k)]^T R(k) [\underline{U}(k) - \underline{U}_d(k)] \tag{5.2}$$

then $\underline{U}(k)$ can be chosen to minimise $J(k)$ subject to (5.1). This leads to the solution

$$\begin{aligned} \underline{U}(k) = {} & R(k)^{-1} A_o^T [A_o R(k)^{-1} A_o^T]^{-1} \underline{u}(k) \\ & + \{I - R(k)^{-1} A_o^T [A_o R(k)^{-1} A_o^T]^{-1} A_o\} \underline{U}_d(k) \end{aligned} \tag{5.3}$$

where the matrix $\{.\}$ is a projection operator, i.e. $\{.\}^2 = \{.\}$.

The solution to the input disaggregation problem within the optimisation framework thus introduces the possibility of using a series of one-step optimisations to counter undesirable short-term effects while employing the simple model to implement the major strands of long-term economic policy. The problem can be resolved within a more general optimisation framework and appears to warrant some further study.

6. ASPECTS OF THE LINK-UP

Before presenting the computational results from the link-up it is worth considering the important differences that exist between the two models and some of the critical assumptions which underpin our approach to controlling the LBS model.

6.1 Differences between PREM and LBS Model

1. (*a*) The LBS econometric model is a nonlinear model estimated mainly on levels data using an OLS estimator;
   (*b*) The PREM model is a linear model estimated on the growth rates of levels data using an FIML estimator on blocks of equations;
2. (*a*) The LBS model has approximately 150 endogenous variables, 200 identities, and 100 exogenous variables, including more than 50 dummy variables;
   (*b*) The PREM model has twenty endogenous variables, 17 identities and 15 exogenous variables (7 'true' exogenous, 8 controls).
3. (*a*) The important dynamics in the LBS model are captured with Almon lags;
   (*b*) The dynamics of the PREM model are captured with rational lags.
4. (*a*) The LBS model is designed largely for forecasting and policy simulation;
   (*b*) The PREM model is used principally for control exercises.
5. (*a*) The LBS model includes
   a. Building Societies
   b. Flow of Funds
   c. Public Sector Borrowing Requirement
   d. Balance of Payments
   e. Interest Rates
   f. Bank Advances/Endogenous Money Supply
   g. Corporate Taxation
   h. World Commodity Prices
   (*b*) The PREM model has none of these.

6.2 Assumptions

(*a*) Policy is determined on the basis of information on economic variables which is assumed complete up to the previous quarter.
(*b*) Measurement error is absent so that data collected initially is not subject to subsequent revision.
(*c*) Once information about the state is available the government is able to react immediately by changing the values of its instruments.

Both assumptions (*a*) and (*c*) mean that in the terminology of stabilisation theory [Ando and Brown (1968)], there is no 'inside' lag.

7. NUMERICAL RESULTS

7.1 Basic Simulation and Control Runs

Phase 1: LBS Model Simulation
Simulate LBS model over period 1970(1)–1974(4) using actual historical inputs for the exogenous variables and 'controls'.

Phase 2: PREM Model Simulation
Set desired trajectories of the exogenous variables and controls of the PREM model equal to historical values (1970(1)–1974(4)) and execute PREM model simulation run. This can be achieved by placing zero weights on the endogenous variables in the performance criterion.

Phase 3: LBS/PREM Link-up
Maintain the desired trajectories of the exogenous variables and controls for the PREM model as in Phase 2; place weights on selected endogenous variables. Constrain tax controls so they only change at budget time. Execute link-up algorithm as described in section (4) above over period 1970(1)–1974(4).

Compare results with the output of Phase 1.

7.2 Choice of Performance Index

The initial choice of performance index for the PREM model was made with the link-up in mind. The primary aim was to exert a downward pressure on the level of unemployment, an upward pressure on the export level and a moderating influence on prices by suitable costing of growth rate variables. Secondly, it was decided to move cautiously away from the simulation case (in which all output weights are zero and there is no feedback) by keeping a careful hold on the instruments.

The choice of (diagonal) weighting matrices for the quadratic objective function and the desired trajectories were as follows:

Endogenous variables	*desired value*	*weight*
% Δ unemployment	−.5 (constant)	3
% Δ exports	1.75 ,,	5
% Δ GDP deflator	1.0 ,,	3

Controls

% Δ current government expenditure	historical data		2
% Δ exchange rate index	" "	.01	
% Δ basic tax rate	" "	30	(.01 : 2nd qtr)
% Δ in aggregate allowances	" "	30	(.01 : 2nd qtr)

All other outputs and controls were screened out by zero-weighting outputs and by placing large weights on the controls.

7.3 Interpretation of Results

When interpreting the results of the link-up we shall focus attention on the main economic aggregates: unemployment, the balance of payments, the rate of inflation and the Public Sector Borrowing Requirement. Government current expenditure, the exchange rate, the basic tax rate and allowances have been used as instruments. In the small model only unemployment, exports, and the GDP deflator among the endogenous variables have been weighted. Since neither the balance of payments in levels nor the public sector borrowing requirement enter directly into the small model, we are deriving the implications of a particular optimal control run for them by using the big model. We are not able to weight these variables explicitly in the performance index for the small model.

The simulation performance of the LBS model is used as a benchmark against which the closed-loop and open-loop results are evaluated. The LBS model simulates the actual performance of the economy during 1970–74, with the exception of the inflation rate, very well. The wages sector of the LBS model is designed specifically to forecast from the middle of 1975 onwards under the various phases of the social contract (and, therefore, is ill-suited to simulation of the 1970–74 period) which gives a downward bias to the inflation figures during this historical period.

7.4 Case I. Closed-loop Control: Full Information about Exogenous Variables

Results for various variables, both outputs and controls are plotted in Figures 10.3 and 4. The results for a more extensive list of variables in both levels and percentage changes over the same period of the previous year are contained in Tables 10.1–7.

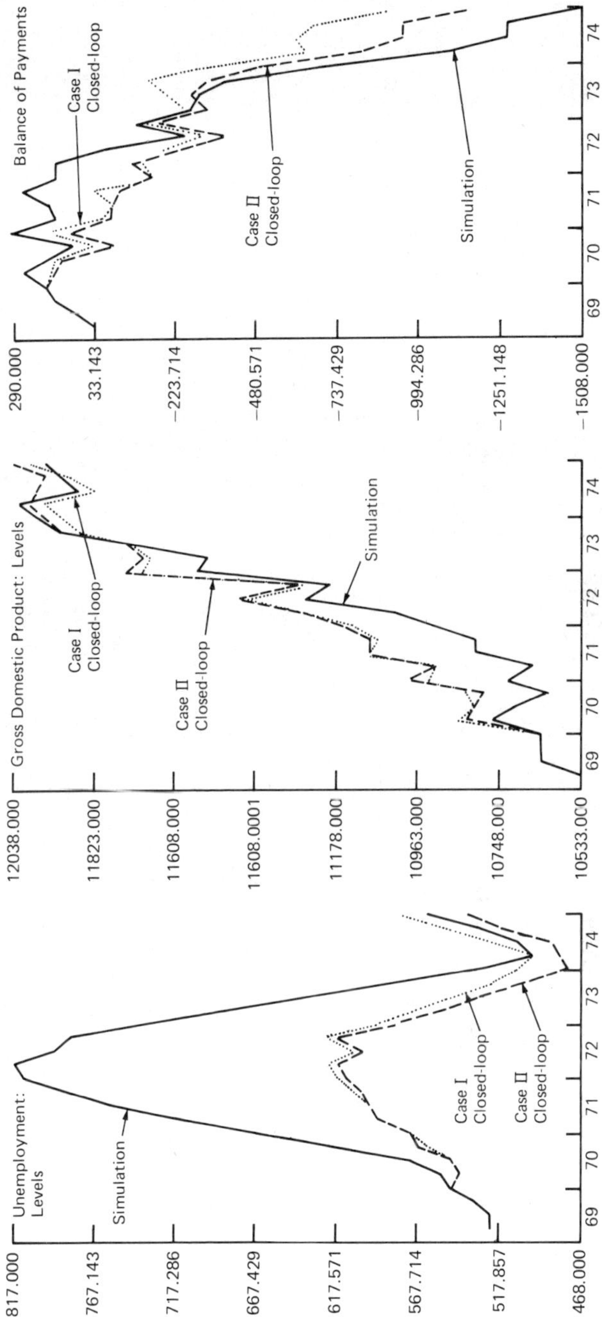
Unemployment: Levels
817.000
767.143
717.286
667.429
617.571
567.714
517.857
468.000
Simulation
Case I Closed-loop
Case II Closed-loop
69
70
71
72
73
74
Gross Domestic Product: Levels
12038.000
11823.000
11608.000
11608.0001
11178.000
10963.000
10748.000
10533.000
Case I Closed-loop
Case II Closed-loop
Simulation
Balance of Payments
290.000
33.143
−223.714
−480.571
−737.429
−994.286
−1251.148
−1508.000
Case I Closed-loop
Case II Closed-loop
Simulation

FIG. 10.3

TABLE 10.1 Gross Domestic Product

In each table '%Δ' denotes percentage change from same time period of previous year.

		Simulation	%Δ	Case I Closed-loop	%Δ	Case II Closed-loop	%Δ	Case III Open-loop	%Δ
1969	1	10532.773	1.243	10532.773	1.243	10532.773	1.243	10532.773	1.243
	2	10620.812	2.601	10629.812	2.601	10629.812	2.601	10629.812	2.601
	3	10629.812	2.282	10629.812	2.282	10629.812	2.282	10629.812	2.282
	4	10640.594	1.753	10640.594	1.753	10640.594	1.753	10640.594	1.753
1970	1	10773.937	2.290	10858.078	3.088	10842.066	2.936	10843.457	2.950
	2	10711.004	0.764	10812.172	1.716	10813.687	1.730	10806.109	1.659
	3	10614.574	−0.143	10828.238	1.867	10801.301	1.613	10725.801	0.903
	4	10716.629	0.715	10947.988	2.889	10986.301	3.249	10835.684	1.833
1971	1	10652.418	−1.128	10909.516	0.474	10907.945	0.608	10774.453	−0.636
	2	10812.316	0.946	11092.316	2.591	11090.012	2.555	10922.094	1.073
	3	10804.574	1.790	11080.016	2.325	11086.086	2.637	10890.223	1.533
	4	10916.039	1.861	11132.508	1.685	11156.734	1.551	10976.504	1.300
1972	1	11033.035	3.573	11272.105	3.324	11254.402	3.176	11071.047	2.753
	2	11265.000	4.187	11417.309	2.930	11434.453	3.106	11294.562	3.410
	3	11200.211	3.662	11258.199	1.608	11277.551	1.727	11205.363	2.894
	4	11549.691	5.805	11700.113	5.099	11744.117	5.265	11545.309	5.182
1973	1	11622.309	5.341	11671.000	3.539	11684.352	3.820	11619.707	4.956
	2	11729.391	4.122	11722.242	2.671	11745.926	2.724	11739.125	3.936
	3	11905.547	6.298	11857.703	5.325	11911.551	5.622	11946.820	6.617
	4	11983.738	3.758	11910.266	1.796	11960.746	1.845	12071.195	4.555
1974	1	12012.914	3.361	11950.043	2.391	11992.707	2.639	12150.383	4.567
	2	11863.586	1.144	11825.992	0.885	11857.977	0.954	12038.668	2.552
	3	11907.937	0.020	11884.797	0.228	11957.953	0.390	12107.656	1.346
	4	11961.500	−0.186	11998.105	0.738	12038.402	0.649	12181.012	0.910

Table 10.1 (*continued*)

	Simulation	%Δ	*Case I Closed-loop*	%Δ	*Case II Closed-loop*	%Δ	*Case III Open-loop*	%Δ
1969	42432.992	1.969	42432.992	1.969	42432.992	1.969	42432.992	1.969
1970	42816.145	0.903	43446.477	2.388	43443.355	2.381	43211.051	1.834
1971	43185.348	0.862	44214.355	1.767	44240.777	1.836	43563.273	0.815
1972	45047.937	4.313	45647.727	3.242	45710.523	3.322	45116.281	3.565
1973	47240.984	4.868	47161.211	3.316	47302.574	3.483	47376.848	5.011
1974	47745.937	1.069	47658.937	1.055	47847.039	1.151	48477.719	2.324

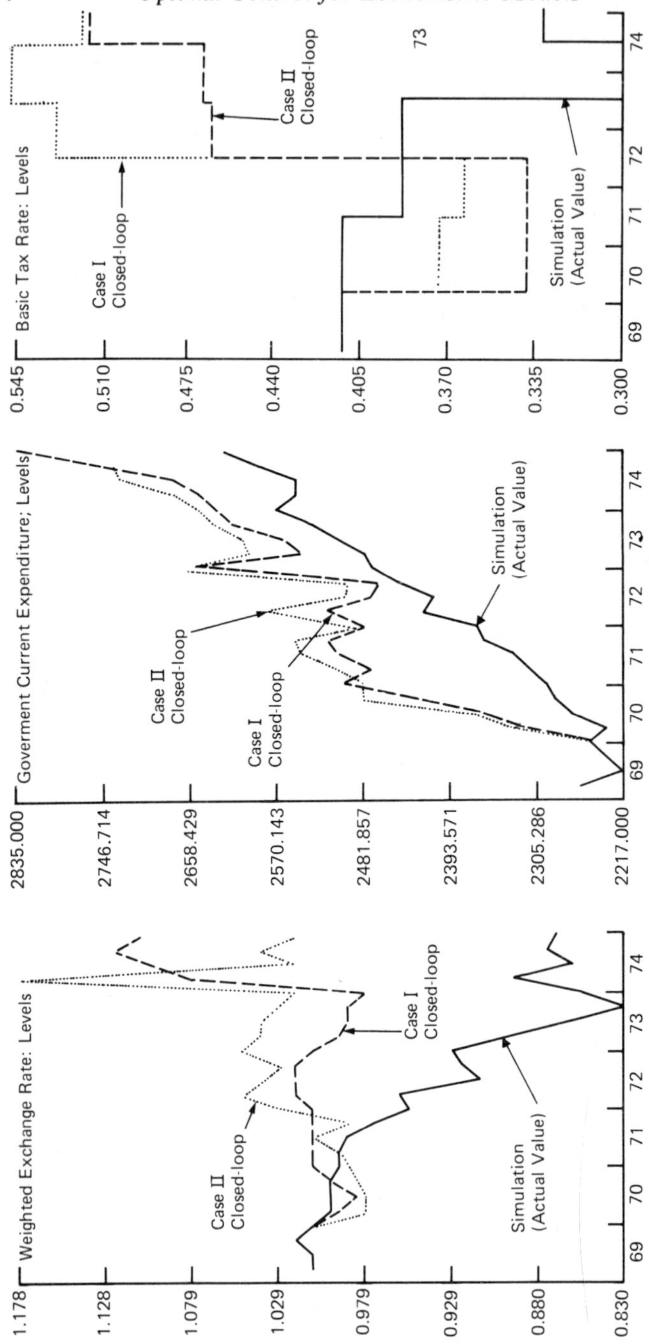
Basic Tax Rate: Levels
0.545
0.510
0.475
0.440
0.405
0.370
0.335
0.300
69
70
71
72
73
74
Case I
Closed-loop
Case II
Closed-loop
Simulation
(Actual Value)
Goverment Current Expenditure; Levels
2835.000
2746.714
2658.429
2570.143
2481.857
2393.571
2305.286
2217.000
Case II
Closed-loop
Case I
Closed-loop
Simulation
(Actual Value)
Weighted Exchange Rate: Levels
1.178
1.128
1.079
1.029
0.979
0.929
0.880
0.830
Case II
Closed-loop
Case I
Closed-loop
Simulation
(Actual Value)

FIG. 10.4

Figure 3a shows the simulated level of unemployment from 1969(1) to 1974(4) (the 1969 data is actual historical data) and the level of unemployment (marked Case I) that results when the closed-loop feedback law calculated from the PREM model is applied to the LBS model. As compared with a simulation result of 817,000 at 1972(1), the closed-loop solution peaks at only 621,000. If we study the behaviour of the controls (Figure 10.4, Tables 10.5–7), it is clear how this is achieved. Historically, the economy was not reflated until the budgets of 1971 and 1972. With the closed-loop control, however, government current

TABLE 10.2 Public sector borrowing requirement and balance of payments

		Public Sector Borrowing Requirement				*Balance of Payments*			
		Simulation	*Case I Closed-loop*	*Case II Closed-loop*	*Case III Open-loop*	*Simulation*	*Case I Closed-loop*	*Case II Closed loop*	*Case III Open-loop*
1969	1	−1198	−1198	−1198	−1198	22	22	22	22
	2	−85	−85	−85	−85	96	96	96	96
	3	230	230	230	230	167	167	167	167
	4	579	579	579	579	175	175	175	175
1970	1	−1131	−1069	−1081	−1052	256	276	252	248
	2	205	455	591	438	191	165	145	157
	3	316	627	727	531	123	33	−9	67
	4	781	1081	1238	984	290	164	104	214
1971	1	−265	27	139	−79	174	16	−31	89
	2	494	711	776	433	196	−6.5	−27	134
	3	719	861	931	635	265	32	−44	180
	4	1090	1173	1265	1015	157	−158	−139	54
1972	1	−758	−615	−589	−835	179	−115	−94	118
	2	737	1.5	282	483	11	−200	−226	−95
	3	685	−76	190	447	−261	−301	−339	−294
	4	1215	615	913	976	−94	−113	−164	−55
1973	1	−377	−1019	−733	−613	−285	−250	−316	−280
	2	1329	199	590	1329	−301	−187	−278	−363
	3	1276	362	781	1311	−381	−155	−325	−499
	4	1486	486	915	1558	−689	−305	−489	−704
1974	1	−455	−1408	−999	−351	−1095	−628	−825	−1126
	2	1882	1492	1429	2274	−1278	−605	−938	−1408
	3	1404	848	908	1787	−1288	−651	−946	−1450
	4	2567	1872	1921	2951	−1508	−902	−1157	−1704
1969		−474	−474	−474	−474	460	460	460	460
1970		171	1094	1475	900	861	643	491	686
1971		2038	2772	3110	2004	793	−116	−241	457
1972		1880	−74	797	1070	−164	−724	−873	−326
1973		3713	29	1553	3585	−1657	−897	−1409	−1846
1974		5398	2804	3257	6661	−5168	−2786	−3866	−5689

expenditure increases markedly, basic tax rate is reduced and allowances are increased. In addition, the exchange rate depreciates throughout the whole of 1970. Consequently the public sector borrowing requirement during 1970 rises from £170 m. to £1094 m, and the surplus on the balance of payments is slightly smaller. During 1971 the basic tax rate is reduced marginally, while allowances are actually cut and government current expenditure grows much more slowly. The balance of payments, not surprisingly, swings into deficit, while the PSBR is £700 m larger.

TABLE 10.3 Unemployment

		Simulation		*Case I Closed loop*		*Case II Closed loop*		*Case III Open loop*	
			%Δ		%Δ		%Δ		%Δ
1969	1	525	−2.60	525	−2.60	525	−2.60	525	−2.60
	2	521	−4.40	521	−4.40	521	−4.40	521	−4.40
	3	532	−1.48	532	−1.48	532	−1.48	532	−1.48
	4	546	2.83	546	2.83	546	2.83	546	2.83
1970	1	553	5.31	541	3.07	543	3.50	543	3.46
	2	575	10.3	546	4.84	549	5.33	550	5.49
	3	621	16.8	561	5.56	568	6.69	579	8.89
	4	665	21.8	571	4.57	572	4.73	607	11.12
1971	1	719	30.0	592	9.34	591	8.72	647	19.14
	2	754	31.3	597	9.29	596	8.59	674	22.70
	3	788	26.8	607	8.12	605	6.67	706	21.95
	4	810	21.8	620	8.56	615	7.49	733	20.73
1972	1	817	13.6	621	4.99	618	4.65	748	15.65
	2	794	5.2	607	1.76	604	1.29	736	9.16
	3	783	−0.68	623	2.56	616	1.79	740	4.46
	4	738	−8.8	593	−4.27	581	−5.52	706	−3.58
1973	1	689	−15.7	566	−8.82	552	−10.76	667	−10.87
	2	640	−19.4	547	−9.92	531	−12.03	624	−15.16
	3	584	−25.4	525	−15.73	503	−18.39	568	−23.00
	4	534	−27.7	507	−14.52	480	−17.36	508	−28.04
1974	1	500	−27.4	498	−12.01	468	−15.25	455	−31.73
	2	508	−20.6	521	−4.72	490	−7.73	440	−29.59
	3	533	−8.6	553	5.46	517	2.75	440	−22.58
	4	563	5.6	577	13.77	539	12.36	445	−12.42
1969		531	−1.44	531	−1.44	531	−1.44	531	−1.44
1970		603	13.71	555	4.50	558	5.06	570	7.29
1971		768	27.20	604	8.81	602	7.85	690	21.14
1972		783	1.97	611	1.22	605	0.50	732	6.10
1973		611	−21.9	536	−12.24	516	−14.60	592	−19.15
1974		526	−13.9	538	0.214	503	−2.52	445	−24.83

During 1972 the brakes are put on. Government expenditure grows even more slowly, allowances are cut again and basic tax rate is increased from 36 per cent to 53 per cent, which reduces real disposable income during the next twelve months and of course consumption. Gross domestic product still grows strongly during 1972 because of strong growth in both exports and investment. An increase in basic tax rate to 53 per cent will be enough to make even the most hardened economist wince. But such a large burden is being thrown onto the basic tax rate because there are no other fiscal instruments included in the feedback law apart from allowances. In a fuller analysis the burden could be spread more evenly between fiscal instruments. At the same time the control law is attempting to offset during 1972 the expansionary effects of various exogenous variables that in a fuller study would be among the controls. The effect of the deflationary measures of 1972 is more or less to eliminate the PSBR [although the balance of payments deteriorates further] largely at the expense of consumption and government current expenditure. Over the first three years of the control period the annual rate of increase of the GDP deflator does not differ greatly from the simulation result.

During 1973 further deflationary action is taken. Basic tax rate is *increased* to 54 per cent but the main deflationary effect is achieved by a large cut in allowances. Most of the deflationary action is taken to offset the effects of the large increase in exports during 1973. The PSBR is in a small surplus during 1973 but the balance of payments worsens slightly. In volume terms exports comfortably cover imports but there has been a steady deterioration in the terms of trade during 1972 and 1973. After peaking in 1972(3) the level of unemployment declines steadily during 1973.

By 1974, when the effects of the oil price rise are beginning to be felt, the LBS analogue of the UK economy is in a much better state than the actual economy at the time. The downturn in investment that began during 1973 is partly offset by increases in government expenditure and by a cut in the basic tax rate and increases in allowances. This proves insufficient, however, to offset the rise in unemployment during 1974.

Despite the glaring anomalies in the use of the tax controls, an overall pattern emerges. The control scheme, based as it is on percentage changes in the variables, has exerted a moderating influence on variations in the target variables. By the close of the control period gross domestic product is not significantly higher than in the simulation case. The higher level of government expenditure is offset by reduced consumption. The cumulated public sector borrowing requirement as a

TABLE 10.4

		GDP DEFLATOR %Δ				IMPORT PRICES %Δ			
		Simulation	*Case I Closed loop*	*Case II Closed loop*	*Case III Open loop*	*Simulation*	*Case I Closed loop*	*Case II Closed loop*	*Case III Open loop*
1969	1	2.70	2.70	2.70	2.70	5.65	5.65	5.65	5.65
	2	3.60	3.60	3.60	3.60	1.32	1.32	1.32	1.32
	3	3.78	3.78	3.78	3.78	−0.8	−0.8	−0.8	−0.8
	4	4.43	4.43	4.43	4.43	3.22	3.22	3.22	3.22
1970	1	6.42	6.20	6.26	6.26	0.76	−2.08	−0.21	−0.22
	2	7.35	6.77	6.87	6.89	7.68	4.97	5.62	6.11
	3	8.74	7.55	7.76	7.95	9.79	7.44	9.81	8.16
	4	10.17	8.34	8.53	9.06	8.58	6.69	8.74	6.88
1971	1	9.3	7.18	7.33	8.16	12.95	14.75	14.37	12.17
	2	9.3	7.25	7.38	8.43	10.40	13.69	13.25	10.16
	3	8.4	6.64	6.71	7.89	8.79	8.16	8.96	9.41
	4	7.2	6.14	6.17	7.18	4.70	13.5	9.08	8.76
1972	1	6.16	6.10	6.07	6.68	2.32	11.05	7.15	5.92
	2	4.21	5.32	5.17	5.21	−0.87	10.79	8.12	7.41
	3	2.66	5.32	5.003	4.12	−1.55	12.08	8.57	4.79
	4	1.11	4.78	4.32	2.82	0.41	7.16	5.50	2.18
1973	1	−0.25	4.02	3.45	1.49	−0.6	7.60	6.04	4.20
	2	−0.78	3.76	3.16	0.82	1.49	9.76	7.49	4.32
	3	−1.10	3.30	2.67	0.23	2.54	13.05	10.08	9.59
	4	−0.74	3.45	2.86	0.20	6.53	12.11	11.89	10.93

TABLE 10.4 (continued)

		GDP DEFLATOR %Δ				IMPORT PRICES %Δ			
		Simulation	Case I Closed loop	Case II Closed loop	Case III Open loop	Simulation	Case I Closed loop	Case II Closed loop	Case III Open loop
1974	1	0.29	4.11	3.55	0.76	22.1	27.98	26.56	23.33
	2	2.65	5.48	5.10	2.63	32.1	21.63	30.68	31.14
	3	6.11	7.34	7.22	5.6	31.6	20.51	27.79	26.70
	4	9.81	9.31	9.59	8.99	31.0	23.68	32.13	30.91
1969		3.63	3.63	3.63	3.63	2.33	2.33	2.33	2.33
1970		8.19	7.23	7.36	7.55	6.70	4.26	5.99	5.23
1971		8.54	6.79	6.89	7.91	9.10	12.42	11.32	10.08
1972		3.50	5.37	5.13	4.68	0.06	10.20	7.32	5.04
1973		−0.71	3.63	3.03	0.68	2.49	10.66	8.91	7.27
1974		4.71	6.58	6.38	4.50	29.29	23.37	29.35	28.10

result of using economic policies generated by the feedback law is just half of the simulation result (£6619 m against £13,195 m) while the cumulated balance of payments deficit is reduced from £5335 m to £3980 m.

One point that can be made, drawn from these rough preliminary results, is that optimal control techniques, based as they are on a global hold over system dynamics, introduces a much-needed rational element into the timing of policy changes. This is extremely important for the overall performance of the economic system. In the simulation case reflationary action was delayed for too long to prevent the very high level of unemployment during 1972. Equally deflationary action, because it

TABLE 10.5 Government current expenditure

		Simulation	%Δ	*Case I Closed-loop*	%Δ	*Case II Closed-loop*	%Δ	*Case III Open-loop*	%Δ
1969	(1)	2258	−1.911	2258	−1.911	2258	−1.911	2258	−1.911
	(2)	2217	−2.162	2217	−2.162	2217	−2.162	2217	−2.162
	(3)	2235	−1.845	2235	−1.845	2235	−1.845	2235	−1.845
	(4)	2250	−0.354	2250	−0.354	2250	−0.354	2250	−0.354
1970	(1)	2237	−0.930	2329	3.142	2310	2.296	2310	2.296
	(2)	2273	2.526	2363	6.602	2360	6.456	2357	6.317
	(3)	2290	2.461	2478	10.872	2432	8.797	2371	6.086
	(4)	2295	2.000	2480	10.230	2498	11.044	2372	5.416
1971	(1)	2310	3.263	2505	7.555	2472	7.014	2381	3.100
	(2)	2334	2.684	2545	7.695	2509	6.317	2400	1.823
	(3)	2355	2.838	2551	2.930	2518	3.569	2404	1.374
	(4)	2367	3.137	2490	0.413	2486	−0.498	2398	1.100
1972	(1)	2421	4.805	2577	2.874	2519	1.913	2441	2.498
	(2)	2408	3.171	2499	−1.805	2475	−1.361	2428	1.178
	(3)	2444	3.779	2502	−1.893	2466	−2.085	2463	2.485
	(4)	2472	4.436	2661	6.835	2655	6.786	2492	3.931
1973	(1)	2486	2.685	2596	0.754	2547	1.100	2515	3.049
	(2)	2512	4.319	2608	4.352	2564	3.597	2556	5.272
	(3)	2539	3.887	2636	5.359	2618	6.187	2611	5.981
	(4)	2572	4.045	2647	−0.518	2629	−0.974	2688	7.863
1974	(1)	2550	2.574	2679	3.195	2646	3.909	2711	7.766
	(2)	2549	1.473	2725	4.496	2674	4.292	2728	6.718
	(3)	2585	1.812	2737	3.817	2764	5.560	2769	6.058
	(4)	2622	1.944	2835	7.093	2835	7.829	2810	4.542
1969		8960	−1.571	8960	−1.571	8960	−1.571	8960	−1.571
1970		9095	1.507	9650	7.706	9600	7.143	9410	5.020
1971		9366	2.980	10091	4.566	9985	4.015	9583	1.841
1972		9745	4.047	10239	1.466	10115	1.295	9825	2.523
1973		10109	3.735	10488	2.427	10358	2.407	10371	5.555
1974		10306	1.949	10976	4.659	10919	5.416	11018	6.242

was delayed until 1973 and 1974, could not moderate the steadily deteriorating position of the balance of payments and Public Sector finances during 1972, 1973 and 1974.

7.5 Case II: Closed-loop Control: Exogenous Variables Set Equal to Mean Values

The robustness of the closed-loop controller can be tested further if it is deprived of full information about the trajectories of the exogenous variables. Therefore, Phase 2 of the procedure was modified by replacing the exogenous variable trajectories by their means:

$$\underline{e}(k) = [2.143, 2.353, 2.393, 1.039, 4.606, -.139, 0], \; k = 1, \ldots, 20$$

This is closer to the true circumstances in which the policy-maker finds himself. There is uncertainty about how the exogenous variables will behave in the future.

The results are given in the various tables and figures. Compared with Case I the outcome, as would be expected, is different. The level of unemployment is actually marginally better, but both the PSBR and the balance of payments are worse than in Case I. The cumulated balance of payments is actually worse even than the simulation run, due mainly to higher prices during 1974 that result from devaluation during 1973.

As would be expected, the closed-loop feedback law is sensitive to the amount of information that is available about the exogenous variables; but overall the performance of the LBS model under control is better than under simulation.

7.6 Case III A Comparison with Open-loop Control

To reiterate the point made in Westcott *et al.* (Chapter 1 above): in a perfectly known and disturbance-free world there would be no advantage in employing our chosen economic policies other than as a sequence of fixed and predetermined values for the various instruments of economic policy. In such circumstances it would be justifiable to rely completely on simulation results obtained from a model in formulating policy for the actual economy. If the PREM model were a perfect representation of the LBS system, then we could use the instrument trajectories generated by optimally controlling the former model to produce an optimal simulation of the LBS model. Thus there would be no purpose in employing control models for generating closed-loop policies rather than open-loop (simulation) policies.

TABLE 10.6

		BASIC TAX RATE				WEIGHTED EXCHANGE RATE INDEX			
		Simulation	*Case I Closed-loop*	*Case II Closed-loop*	*Case III Open-loop*	*Simulation*	*Case I Closed-loop*	*Case II Closed-loop*	*Case II Open-loop*
1969	1	0.412	0.412	0.412	0.412	1.010	1.010	1.010	1.010
	2	0.412	0.412	0.412	0.412	1.011	1.011	1.011	1.011
	3	0.412	0.412	0.412	0.412	1.015	1.015	1.015	1.015
	4	0.412	0.412	0.412	0.412	1.005	1.005	1.005	1.005
1970	1	0.412	0.412	0.412	0.412	1.004	0.981	0.996	0.996
	2	0.412	0.372	0.338	0.384	1.000	0.981	0.985	0.989
	3	0.412	0.372	0.338	0.384	0.997	0.984	1.003	0.988
	4	0.412	0.372	0.338	0.384	0.996	0.991	1.007	0.988
1971	1	0.412	0.372	0.338	0.384	0.994	0.996	1.010	0.986
	2	0.387	0.364	0.339	0.405	0.990	1.010	1.011	0.982
	3	0.387	0.364	0.339	0.405	0.975	0.990	1.018	0.981
	4	0.387	0.364	0.339	0.405	0.952	1.030	1.006	0.982
1972	1	0.387	0.364	0.339	0.405	0.958	1.046	1.020	0.981
	2	0.387	0.529	0.465	0.455	0.916	1.043	1.016	0.978
	3	0.387	0.529	0.465	0.455	0.922	1.026	1.022	0.973
	4	0.387	0.529	0.465	0.455	0.929	1.052	1.012	0.962
1973	1	0.387	0.529	0.465	0.455	0.895	1.035	0.995	0.948
	2	0.300	0.545	0.467	0.367	0.861	1.043	0.987	0.942
	3	0.300	0.545	0.467	0.367	0.830	1.027	0.990	0.935
	4	0.300	0.545	0.467	0.367	0.854	1.024	0.977	0.924
1974	1	0.300	0.545	0.467	0.367	0.894	1.178	1.078	0.981
	2	0.330	0.518	0.515	0.354	0.862	1.019	1.103	0.928
	3	0.330	0.518	0.515	0.354	0.877	1.037	1.122	0.995
	4	0.330	0.518	0.515	0.354	0.869	1.023	1.107	0.985

TABLE 10.6 *(continued)*

	BASIC TAX RATE				WEIGHTED EXCHANGE RATE INDEX			
	Simulation	*Case I Closed-loop*	*Case II Closed-loop*	*Case III Open-loop*	*Simulation*	*Case I Closed-loop*	*Case II Closed-loop*	*Case II Open-loop*
1969	1.648	1.648	1.648	1.648	1.010	1.010	1.010	1.010
1970	1.648	1.529	1.427	1.564	0.999	0.984	0.998	0.990
1971	1.573	1.464	1.354	1.598	0.978	1.007	1.011	0.982
1972	1.548	1.950	1.735	1.770	0.931	1.042	1.017	0.973
1973	1.282	2.164	1.868	1.556	0.860	1.032	0.987	0.937
1974	1.290	2.100	2.013	1.430	0.875	1.064	1.103	0.985

TABLE 10.7

		EARNED INCOME ALLOWANCE				AGGREGATED PERSONAL ALLOWANCES			
		Simulation	*Case I Closed-loop*	*Case II Closed-loop*	*Case III Open-loop*	*Simulation*	*Case I Closed-loop*	*Case II Closed-loop*	*Case III Open-loop*
1969	1	1479.6	1479.6	1479.6	1479.6	1708.6	1708.6	1708.6	1708.6
	2	1496.1	1496.1	1496.1	1496.1	1906.7	1906.7	1906.7	1906.7
	3	1518.1	1518.1	1518.1	1518.1	1906.7	1906.7	1906.7	1906.7
	4	1565.6	1565.6	1565.6	1565.6	1906.7	1906.7	1906.7	1906.7
1970	1	1595.2	1579.3	1579.3	1751.2	1906.7	1923.4	1923.4	1751.2
	2	1675.9	1954.1	2053.2	2129.7	2404.1	2379.8	2500.5	2129.7
	3	1728.6	1979.5	2079.8	2157.2	2404.1	2410.8	2532.9	2157.2
	4	1745.4	1987.5	2088.4	2165.9	2404.1	2420.5	2543.4	2165.9
1971	1	1792.2	2009.8	2111.7	2190.3	2404.1	2447.7	2571.8	2190.3
	2	1836.0	1932.8	1991.8	1978.7	2318.2	2353.9	2425.8	1978.7
	3	1891.2	1958.2	2018.1	2005.0	2318.2	2384.9	2457.8	2005.0
	4	1922.6	1972.7	2033.1	2019.7	2315.2	2402.5	2476.0	2019.7
1972	1	1972.1	1995.5	2056.5	2043.4	2318.2	2430.3	2504.5	2043.4
	2	2050.5	1727.6	1903.2	2275.3	3129.6	2104.0	2317.8	2275.3
	3	2108.5	1746.6	1924.1	2300.7	3129.6	2127.2	2343.3	2300.7
	4	2192.3	1775.1	1955.5	2337.4	3129.6	2161.8	2381.5	2337.4
1973	1	2276.5	1802.7	1985.9	2374.3	3129.6	2195.5	2418.6	2374.3
	2	0	1000.4	1141.9	1720.3	4044.5	1218.4	1390.6	1720.4
	3	0	1000.3	1141.8	1720.3	4044.5	1218.2	1390.6	1720.4
	4	0	1000.2	1141.7	1720.3	4044.5	1218.1	1390.5	1720.4
1974	1	0	1000.3	1141.8	1720.3	4044.5	1218.2	1390.5	1720.4
	2	0	1205.6	1245.2	2054.9	4417.9	1468.3	1516.5	2054.9
	3	0	1205.6	1245.3	2054.9	4417.9	1468.2	1516.6	2054.9
	4	0	1205.7	1245.4	2054.9	4417.9	1468.4	1516.7	2054.9

TABLE 10.7 (*continued*)

	EARNED INCOME ALLOWANCE				AGGREGATED PERSONAL ALLOWANCES			
	Simulation	*Case I Closed-loop*	*Case II Closed-loop*	*Case III Open-loop*	*Simulation*	*Case I Closed-loop*	*Case II Closed-loop*	*Case III Open-loop*
1969	6059.4	6059.4	6059.4	6059.4	7428.7	7428.7	7428.7	7428.7
1970	6745.1	7500.3	7800.6	8204.0	9119.0	9134.5	9500.1	8204.0
1971	7442.0	7873.5	8154.7	8193.7	9353.7	9588.9	9931.4	8193.7
1972	8323.4	7244.8	7839.3	8956.8	11707.0	8823.2	9547.2	8956.8
1973	2276.5	4803.6	5411.3	7535.2	15262.9	5850.2	6590.3	7535.3
1974	0	4617.2	4877.7	7885.0	17298.1	5623.1	5940.4	7885.0

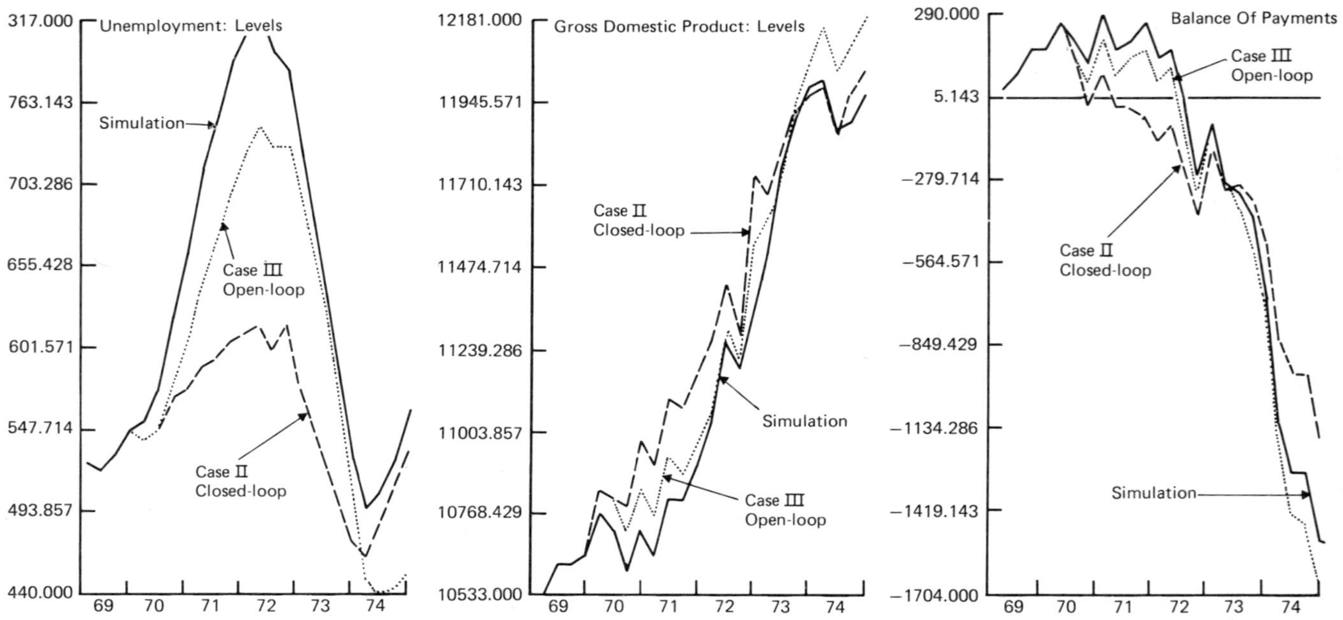
317.000
763.143
703.286
655.428
601.571
547.714
493.857
440.000
Unemployment: Levels
Simulation
Case III
Open-loop
Case II
Closed-loop
69
70
71
72
73
74
12181.000
11945.571
11710.143
11474.714
11239.286
11003.857
10768.429
10533.000
Gross Domestic Product: Levels
Case II
Closed-loop
Simulation
Case III
Open-loop
69
70
71
72
73
74
290.000
5.143
−279.714
−564.571
−849.429
−1134.286
−1419.143
−1704.000
Balance Of Payments
Case III
Open-loop
Case II
Closed-loop
Simulation
69
70
71
72
73
74

FIG. 10.5

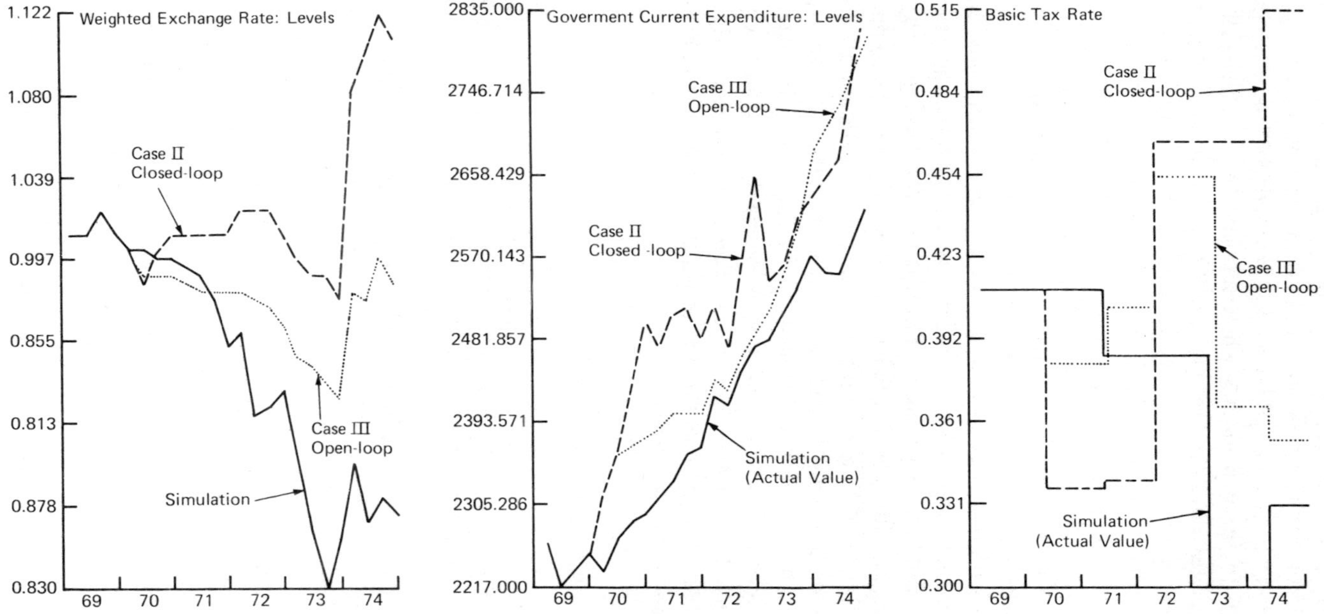
Weighted Exchange Rate: Levels
1.122
1.080
1.039
0.997
0.855
0.813
0.878
0.830
69
70
71
72
73
74
Case II
Closed-loop
Case III
Open-loop
Simulation
Goverment Current Expenditure: Levels
2835.000
2746.714
2658.429
2570.143
2481.857
2393.571
2305.286
2217.000
Case III
Open-loop
Case II
Closed -loop
Simulation
(Actual Value)
Basic Tax Rate
0.515
0.484
0.454
0.423
0.392
0.361
0.331
0.300
Case II
Closed-loop
Case III
Open-loop
Simulation
(Actual Value)

FIG. 10.6

The point need not be laboured that, in reality, there are substantial differences between the PREM and LBS models. Hence the same inputs applied to the two models will yield differences in the values of the endogenous variables. However the PREM model is used in the link-up as a means of reconstructing (approximately) certain key parts of the LBS model's state vector by using the LBS model outputs. Therefore the feedback law operates on an approximation to the LBS state, in order to produce the instrument values, rather than on the PREM state. Thus a self-correcting effect is introduced by the feedback which compensates somewhat for the model mismatch.

The LBS model was run under open-loop control using the PREM instrument trajectories generated in Case II. The results are shown in the Tables and Figures 10.5 and 6. The LBS model behaves in a way very like the original simulation. The level of unemployment (as compared with Case II) is much higher, while the cumulated PSBR and the balance of payments deficit are much larger than even in the LBS simulation case.

Overall the cyclical movements in the main economic magnitudes is little different from the simulation. Applying the control as a closed-loop feedback law, rather than in an open-loop mode, clearly makes a significant difference to the way in which the LBS model behaves over the chosen control period.

8. CONCLUSIONS

Preliminary results indicate that employing a control model for policy formulation can lead to improvements in performance even when future exogenous information is lacking and there is some model mis-specification.

The deterioration of performance when open-loop policies are used suggest that policy formulation via simulation alone is not making the best of the available system information. However, at this point in time, we note that further control exercises are necessary before any firmer conclusions can be drawn from the link-up about the sort of detailed improvements that can be made to traditional policy making methods.

Acknowledgements

The financial support of the SSRC for the Programme of Research into Econometric Methods is gratefully acknowledged. We would also like to thank John Eaton of the London Business School for many helpful suggestions and Peter Warburton and Jan Laury who patiently answered many questions about the LBS econometric model.

REFERENCES

Ando, E. and Brown, E. C. (1968), 'Lags in Fiscal Policy – A Summary', pp. 308–12, in Smith, W. L., and Teigen, R. L. (eds), *Readings in Money, National Income and Stabilisation Policy* (Richard D. Irwin Inc.).

Aoki, M. (1971), 'Aggregation', Ch. 5 in *Optimisation Methods for Large-Scale Systems*, ed. D. A. Wismer (McGraw-Hill).

Bray, J. (1975), 'Optimal Control of a Noisy Economy with the UK as an Example', *J. R. Stat. Soc.* (A), 138.

Chidambara, M. R. (1969), 'Two Simple Techniques for the Simplification of Large Dynamic Systems', *Proc. JACC*, pp. 669–74.

Preston, A. J., and Wall, K. D. (1973a), 'An Extended Identification Problem for State Space Representations of Econometric Models', PREM Discussion Paper no. 6.

Preston, A. J. and Wall, K. D. (1973b), 'Some Aspects of the Use of State Space Models in Econometrics', PREM Discussion Paper no. 5.

Shamash, Y. (1975), 'Linear System Reduction Using Pade Approximation to Allow Retention of Dominant Modes', *Int. J. Control*, vol. 21, no. 2, pp. 257–72.

Wall, K. D. (1976), 'FIML Estimation of Rational Distributed Lag Structural Form Models', PREM Discussion Paper no. 11.

Wall, K. D., and Westcott, J. H. (1975), 'Policy Optimisation Studies with a Simple Control Model of the U.K. Economy', IFAC/75 Congress, Boston, Mass., 24–30 Aug.

Appendix

As far as possible, we have not tampered with the specification of the LBS model, except for some cosmetic redefinitions of endogenous and exogenous variables to ensure that the exogenous variables and controls in the PREM model were also exogenous in the LBS model. This meant exogenising in the LBS model world exports, other personal incomes, government current grants to the personal sector, national insurance contributions, and subsidies.

One major change, however, was necessary. The LBS model has a very detailed and disaggregated tax model: Dorrington and Renton (1975). If we had data on every individual's personal circumstances, it would be possible to obtain the aggregate tax liability by simple addition. Since this is not possible, the LBS has divided all taxpayers into 25 income classes. Each case is then subdivided according to a number of further characteristics, such as marital status, source of income, number of dependants and so on. Despite a number of simplifications this gives approximately 2600 subgroups within each income class, giving over 60,000 groups in total. The model allows the effect on aggregate tax liability and the distribution of this liability over income classes to be gauged as the tax instruments (controls), net incomes, or various demographic characteristics alter. The tax model can be used to generate quarterly estimates of tax liability, given estimates of income flows. But because of its complexity, a summary function for aggregate tax liability is derived and grafted on to the actual econometric model.

Liability to income tax per tax unit can be represented as:

$$TLBN = \alpha_0 + \beta_1 (YNE/NUT) + \beta_2 (YNI/NUT) \qquad \text{(A.1)}$$

where

$TLBN$ = income tax liability per tax unit
YNE = net earned income
YNI = net unearned income
NUT = number of tax units.

A similar function can be written for surtax liability. Allowance is also made for administrative delays in the collection of taxes. Of course, as should be clear from the preceding discussion, the coefficients themselves are in general time-varying; and are themselves dependent upon demographic factors, and tax instruments. These coefficients are, therefore, estimated within the tax model and can appear as exogenous inputs in the econometric model. The econometric model and the tax model can also be run in tandem so that the feedback effects of various tax changes on income flows can be allowed for. The version of the LBS model (August 1975 version) that at present is available to PREM only has the tax coefficients as exogenous variables.

The total personal tax yield is finally obtained by applying the basic tax rate to dividends and the composite tax rate to building societies' payment of interest to the personal sector.

For the purposes of a control exercise, very few of the tax instruments actually appear in the LBS model directly; their influence is felt via the tax model. Thus we have the problem of mapping the controls in the PREM model (Holly *et al.* (1977)) into the equivalent controls in the LBS model.

The problem was solved by replacing the personal tax equations in the LBS model by an aggregate function on the lines of equation 17 in the PREM model; but expressed in levels estimated using OLS, and purged of as many inconsistencies between Inland Revenue and National Income data as possible.

Inland revenue data on the tax yield of the personal sector contains elements besides persons. But taxes paid by life assurance companies is probably the only important element. Taxes paid on government grants are also removed; so that the dependent variable is now:

$$TYJ - TLFA - TGG \qquad \text{(A.2)}$$

where

$TLFA$ = taxes paid by life assurance companies
TGG = taxes paid on government grants
TYJ = taxes paid on personal income.

The definition of taxable income is more problematic.

Other personal income has to be adjusted to exclude life assurance companies' receipts of interest and dividends. But this is not a straightforward matter. Define

$$YRDJ = YDIJ + YRJ - ORNT + EIP \qquad \text{(A.3)}$$

where

$YDIJ$ = personal income from dividends and interest
$ORNT$ = Imputed rent of owner-occupiers
EIP = Personal interest payments
YRJ = Personal income from rent.

If it is assumed that receipts of interest and dividends by life assurance companies ($YDIL$) is some constant proportion of $YRDG$, *Gross taxable other personal income* is defined as:

$$YJOT = YJO - ORNT - ECCH - YDIL + EIP \qquad \text{(A.4)}$$

where

YJO = other personal income
$ECCH$ = current transfers to charities from companies (free from tax).

Finally, Taxable Income is defined to be:

$$YJT = YJ - [(YEC + YJG) + (YJO - YJOT) + LZRI + (EIP - \alpha LZRI)^{*} DUMM + YJOT - A(L)JYOT] \qquad \text{(A.5)}$$

where

YEC = employers' contributions
YJG = income from current government grants.

Both these items, it is assumed, are not liable to tax (taxes paid on YJG having already been netted out).

The second component in parenthesis ($YJO - YJOT$) represents non-taxable other income.

$LZRI$ = building societies' receipts for payments of interest on mortgage advances.

The treatment of interest payments to building societies is slightly different from other interest payments as the latter may sometimes be deductible for tax payments. The dummy variable $DUMM$ isolates those periods when it has been possible to claim allowances.

$YJOT$, taxable other income, is introduced with the lag profile:

$$A(L) = 0.05 + 0.20B + 0.5B^2 + 0.2B^3 + 0.05B^4 \qquad \text{(A.6)}$$

as this form of income may not be assessed and tax not paid until some delay has occurred.

Two of the series used, $TLFA$ and EIP, were available only on an annual basis. A quarterly series for $TLFA$ was obtained by interpolation. EIP, however, had to be treated in a slightly different way. An annual regression was run to explain EIP and resulted in:

$$EIP = \underset{(16.4)}{1.25078}\, LZRI + \underset{(6.7)}{1.0384}\, EBI \tag{A.7}$$

$\bar{R}^2 = .996$
$SE = 49.5$

EBI is a proxy for payments of bank interest.

$$EBI = \frac{(RLB + 27.5)(KBMS + KBMS_{-1})}{8000} \tag{A.8}$$

where

$$RLB = \text{minimum lending rate}$$
$$KBMS = \text{stock of bank advances.}$$

The coefficient α in the identity for taxable income is set equal to the coefficient on LZRI in the equation above.

The quarterly series for EIP was constructed by using the above equation to obtain 'calculated' values from quarterly values of the independent variables and then adding the interpolated residuals from the annual regression.

Allowances
Only the major allowances are taken into account: the single, married, child and wife's earned income allowance. The actual allowance is multiplied by an estimate of the number of claims in each category. The allowances were then aggregated into a single variable. In addition, an earned income allowance was introduced amounting to two-ninths of total earned income up to 1973(1) and zero thereafter. Earned income is defined as wages and salaries, Forces pay and self-employment income.

The finally estimated equation was

$$\frac{(TYJ - TLFA - TGG)}{TRY} = -1{,}195.92 + 0.77346\, YJT - 0.25265\, MSCR - .6913\, EARN \tag{A.9}$$

$\bar{R}^2 = .996$
$SE = £120.2$ m.
$DW = 1.4$
1967(2) – 1974(4)

where

$$MSCR = \text{personal allowances}$$
$$EARN = \text{earned income relief}$$
$$TRY = \text{basic tax rate.}$$

The simulation performance of the LBS model was compared before and after the substitution of this aggregate relationship and no serious divergences of performance were found. The results are available for inspection.

In the control exercises, the aggregated allowances in the PREM model were allocated equally between the earned income allowance and the personal allowances.

REFERENCE

Dorrington, J. C., and Renton, G. A., *Personal Income Taxation*, LBS Discussion Paper no. 33 (Sep 1975).

The idea and the estimated equations for simplifying the personal tax sector of the LBS model were provided by Mark Salmon (now at the Australian National University), who developed them while at the Bank of England.

11 Managing the Modern Economy: Econometric Specification

Lawrence R. Klein

THE ROLE OF MODELS

The legacy of the Keynesian Revolution is a method or an approach to macroeconomic management that attempts largely, but not entirely, to manipulate the level of demand through fiscal and monetary policy instruments in order to achieve economic stabilisation. This has been assumed to mean steady operation of the economy at or near full employment. In the open economy, this approach must be extended to include manipulation of trade policy instruments.

I do not want to belittle the enormous breakthrough in our thinking about the modern economy that followed the development of these ideas, nor do I fail to recognise the relatively successful operation of Keynesian-inspired policies throughout the Western world. I would suggest, however, that demand management policies are not enough to cope with the range of problems facing the modern economy and shall try to describe in this paper what I regard as a more substantial approach to economic management.

The Keynesian analysis of both macro theory and policy reached what I would call its highest stage of development in its presentation in formal mathematical-statistical models of the economy. The original concept was not in terms of formal models, it was essentially a 'cut-and-try' method by which a highly selective number of exogenous variables or parameters, under control of policy-making bodies, would be manipulated until the right outcome was achieved. This might be called 'fine-tuning' and was practised to a large extent by the economic advisers to the Kennedy-Johnson administrations in the United States during the

1960s. It achieved notable success with the investment incentive tax schemes of 1962 and the income tax cut of 1964. The tax surcharge of 1968 did not work out so well for the concept of fine-tuning.

Demand management policies could be implemented with or without formal forecasts and with or without formal models. The formal forecasts could be dispensed with, if policy decisions were made frequently with repeated adjustments to take care of errors in judgement about the contemporary position of the economy. A marked disadvantage to this approach is that economic policy may become highly volatile where a steady policy would be more desirable. If forecasting is taken lightly and policy adjustments are frequently used to override careless estimates about the future of the economy, the public will lack confidence in the policy process and this will work against it.

As for the avoidance of model building, that may be possible as long as the general framework is very simple and contains very few policy instruments. Even within the limited scope of a strongly demand-oriented model, there are severe limits to attempting to formulate policy on the basis of a 2-equation *IS-LM* system.

$$I(r,y) = S(r,y)$$
$$M = L(r,y)$$

where

I = investment function
r = interest rate
M = (exogenous) money supply
S = Savings function
y = income level

The modern large-scale model is a substantial amplification of this system. The Wharton Model of some 400 equations has been labelled as an IS-LM construct.[1] It would not be feasible to make policy in terms of the detailed instruments contained in this system without having a formal model at hand. But the case for model building is even stronger if we extend our choices to the supply side as well as to traditional demand management.

In a subtle way, supply factors are implicitly taken into account in the macroeconomic Keynesian model if the *IS-LM* construct is extended to include price determination, a production function, labour supply and labour demand. Factor *demand* for labour and capital actually deals with conditions of *supply* because the factors of production are demanded primarily in order to be used in supplying goods, as shown by the production function. There is, however, more demand than supply detail in such a system and the main forces working toward determination of GNP in a system solution are operative from the demand side. This became strikingly apparent in attempting to forecast and make policy suggestions in connection with the imposition of the oil embargo in 1973. In that situation, demand-oriented systems of analysis proved to be inadequate in dealing with the problem at hand, and resort to large-scale supply-oriented systems became mandatory.[2] It is in this context that the role of a model becomes evident. The basic tool for obtaining an estimate of the impact of the oil embargo was a large (dynamic, nonlinear) model combining an input-output system with a detailed macro model of final demand, factor incomes, and market clearing for price determination. Models constructed along these lines, especially in the input-output component, feature the conditions of supply, but they are large and complicated to the point that the kind of analysis undertaken with them could not be readily duplicated by informal, judgmental reasoning.

The oil embargo or longer-run energy problem is one of many new issues facing the modern economy and requiring a new appraisal of model-building for policy formation. The new generation of models will be concerned as before with full employment and effective demand, but it will also be concerned with inflation, capacity ceilings, environmental protection, exchange rate stabilisation, credit supplies, energy balance, labour force, structural (long-term) unemployment, food stocks and similar supply-side problems. This latter group of problems, like the embargo of 1973, cannot be well handled with the conventional modes of reasoning in macroeconomics. What is wanted is a combination of the modelling systems inspired by Leontief and Keynes.

The typical or standard macroeconometric model could show the impact of restrained consumer spending, limitations on exports or imports, but it cannot show how much inventory accumulation will be affected by supply shortages or by how much intermediate deliveries of essential materials will affect the supply of goods to the economy. There is an implicit curve or function of aggregate supply implied by the typical macro model, but shifts in such a curve cannot be readily interpreted in terms of specific restrictions in individual sectors of the economy.

The addition of demand detail to the typical or standard macroeconometric model can be implemented through a disaggregation of types of final demand. Instead of a total consumption function

$$C = f(Y)$$

we can estimate and use separate consumption functions

$$C_i = f_i(Y, p_1, \ldots, p_n) \; i = 1, 2, \ldots n$$

$$C = \sum_i C_i$$

C_i = ith type of consumption (food, fuel, shelter, clothing, services, etc.)
Y = real disposable income
p_i = ith price of consumer goods/services

The disaggregation is not quite straightforward in the sense that the spectrum of relative prices must be introduced in specific demand functions, by type, where they may not be relevant variables in the overall function for total consumption (demand).

In the same way, types of investment, types of imports, types of exports, types of public spending, etc., can be introduced without too much trouble in the disaggregated demand-oriented system. Such systems pay a great deal of attention to specific components of demand. To disaggregate the conditions of supply, it is important to give separate treatment to the individual production functions

$$X_i = F_i(K_i, L_i, E_i, M_i) \; i = 1, 2, \ldots, m$$

X_i = ith sector's gross output
K_i = ith sector's capital stock
L_i = ith sector's employment
E_i = ith sector's energy input
M_i = ith sector's material input

For each sector, there will be an accounting identity

$$\Pi_i + rq_iK_i + w_iL_i = p_iX_i - s_iE_i - u_iM_i = (VA)_i$$

$$\sum_i (VA)_i = \text{GNP}$$

Π_i = ith sector's profit
rq_iK_i = ith sector's interest payments
w_iL_i = ith sector's wage payments
p_iX_i = ith sector's gross output value
s_iE_i = ith sector's energy input costs
u_iM_i = ith sector's material input costs

Factor demand and factor price determination equations of the standard (detailed) macro model will generate values for q_i, K_i, w_i, L_i; but there will not be equations for s_i, E_i, or M_i in the demand-oriented model. We can express E_i and M_i in terms of input-output flows as

$$s_i E_i + u_i M_i = \sum_{j=1}^{m} p_j X_{ji}$$

$$s_i E_i = \sum_{1}' p_j X_{ji};\ u_i M_i = \sum_{j}'' p_j X_{ji}$$

X_{ji} = jth sector's output used as input in ith sector. The truncated sums Σ' and Σ'' are assumed to run over energy and material inputs, respectively. In each producing sector, there will also be separate price equations, expressed as mark-ups over unit costs and excess demand (or supply) conditions

$$p_i = g_i\left[\frac{p_i X_i - \Pi_i}{X_i},\ X_i/X_i^c\right]$$

X_i^c = ith sector's capacity output.

All these supply-side equations will be parameterised in highly specific ways, as well as dynamised. A complete input-output system will, however, be needed to implement such a model; that is to say, the bilateral inter-industry flows X_{ji} must be determined. The sector production functions with both original and intermediate factor inputs, together with price equations, are very much supply details. There are so many details, and the system comprising the input-output component must necessarily be so large that it is unthinkable to try to handle such analysis in any way except through formal models. In a formal model of this type, the new types of policy issues can be examined. It represents a major departure in underlying economic philosophy and methodology from the familiar schemes of macro demand management. This is the departure point for my recommendation on managing the modern economy.

DATA PROBLEMS

Models of any type imply substantial data requirements, but the enlarged supply-oriented system imposes even severer data requirements. Inputs, prices, and factor incomes by sector are needed, as shown by the equations set out above. The larger and more detailed is the input-output system, the larger and more detailed must be the final demand system. Although the regular data produced in a national income and product account (NIPA) system are reasonably available and accurate for limited degrees of disaggregation, it is not likely that we shall find enough data readily available for the more detailed final demand system that will be needed to fit well in the enlarged model being considered here.

The particular areas where data are poor or lacking are

inventory stocks by end-use, type and sector
fixed capital stocks by sector
service (invisible) trade items (exports or imports)
prices of internationally traded goods
income (profit and wages) of small enterprises in all sectors
inflation adjustments (inventory and depreciation valuations).

Data are most plentiful and accurate on an annual basis. In many countries, they are available on a quarterly basis, though on a less comprehensive scale. In a few cases, monthly data are available. Many individual statistics are available monthly but not in the form of an interrelated accounting system. The accounting balances are fundamental. Except for the shortest extrapolation horizons, say up to three months at most, it is imperative to have control over accounting balance in a double-entry NIPA system in order to be sure that various statistical series are mutually consistent with one another.

It will be some time before many countries have complete NIPA systems on a monthly basis, yet the problem of management and control of the modern economy depends to a large extent on frequent readings at monthly intervals. The many monthly economic series available can be used as indicators or proxy variables to interpolate or estimate monthly entries in a complete accounting system. A first attempt to construct a monthly model along the specification lines of mainstream econometric models was undertaken by the late T. C. Liu.[3] He managed to make such a system operational, but it has not been maintained. It exists as an interesting research venture.

A related but separate data problem concerns seasonal adjustment. In countries with severe annual weather differences, and this includes most of the major industrial economies of the world, the seasonal swings in economic activity and related magnitudes are fully as large as cyclical or other swings. One approach is to use only unadjusted data in model construction, but to include seasonal indicators as exogenous variables in the model. The trouble with this approach is that the final estimates are best determined in terms of unadjusted values, and users of econometric results are generally at a loss in knowing how to interpret these data. The accounting identities of an economic system actually hold in unadjusted data; they are 'forced' in adjusted data. Until policy-makers and the public at large can be educated to think in terms of unadjusted data, it seems expedient to develop systems that generate seasonally adjusted series. This other approach is the dominant one at the present time – most attempts at using unadjusted data having been dropped. The trouble with the adjusted data is that they contain inaccuracies that are introduced by the process of adjustment. Ever so many reports have been released in recent years on the American economy in terms of seasonally adjusted data that seem to be counter-intuitive. This must remain as a source of data weakness.

In addition to data accuracy or quality, there is a serious problem of quantity. It must be firmly realised that macroeconometric models are based on small samples. In the best of circumstances, if we are to meet the data needs of complete social accounting systems with enough detail to implement policy and other standard model uses, there are about 30 years of data. These can be stretched into more than 100 quarters or more than 350 months, but they still represent only 30 years and cannot, therefore, span many generations of economic behaviour. In a formal sense, serial correlation of semi-annual, quarterly, or monthly data reduce *effective* sample size; so in all cases, we are dealing with small-sample theory about which our statistical procedures are comparatively unknown. Some isolated statistical series may have a longer history, but the whole batch do not.

These data limitations are especially serious for long-run analysis. It is tenuous to try to draw long-run conclusions from short-run data bases, even though it is possible, in principle, to 'solve' a dynamic system far into the future – much further forward than the data base extends backwards. To be very practical, the information content of our statistical samples about fundamental economic behaviour and associated technology/institutional structure enables us to build models that can, at best, be used for careful extrapolations one or two years ahead. We are stretching the limits of our ability to analyse horizons of 5–10 years, although it can be done. For some problems, it is absolutely essential to look ahead for two or three decades. Here, it must be at one's own risk. There is trend-informational content in existing models, but it is pitifully inadequate. Fascinating scenarios can be made ahead for several decades in terms of existing models, but the data base ought to have as much as 50 or more years of information in order to capture the full flavour of changes that we are trying to study in 30-year simulations.

There is a tendency for specialisation these days, and often econometric model builders, programmers, and others using formal models have not prepared their own data bases. They have taken them in prefabricated form from 'data-banks'. I doubt that the analyst using a ready-prepared data file really appreciates the meaning of the informational content of the series that he generates from a model solution. Our problems are so delicate, so sensitive, so error-prone that one who does not fully appreciate the weaknesses from the first step (data preparation) to the last (model application) is likely to commit some terrible blunders that can be truly harmful to society.

MODEL PRECISION

The beauty of the model approach to economic management is that we have a basis for measuring some of the errors involved in decision-making. Sampling error of coefficient determination and residual error associated with each relationship can be estimated from data. The relative sizes of the signal (systematic model information) and the noise (residual deviation of model from actual observation) are, in a sense, known, and the fact is that in economic and social behaviour the noise component is large – larger than in scientific-engineering systems.

It is helpful and instructive to have econometric information that the noise component of a model is large because it tells how much uncertainty should surround the implementation of economic policy. It is helpful to be trying (and succeeding) in ever-present research efforts to reduce the noise component, but its variance is a parameter and not something that can readily be reduced. Some subcomponents can be reduced by effectively increasing sample size (or quality). Similarly, improved model specification may enable us to reduce error. By being able to measure some hitherto unmeasured variables, we may be able to shift some subcomponents away from noise to measurable variables. Other model improvements may help in different ways. Nevertheless, there are severe limits to accuracy improvement. If we measure error by deviation of model generated variables from observed data variables, we are limited in improvement by the very fact that observed data are not exact. The error of measurement of major economic magnitudes may be subject to as much error, by order of magnitude, as we attribute to the noise component. Some economists feel that we may have reached, already, the asymptotic limits of accuracy in model application.[4] This kind of error is inherent in the economic process and is not a property of econometric modelling as such; it is just that the econometric model approach forces and enables us to measure error objectively.

There are different methods of measuring error. We might derive formulas (or experimental numerical approximations) for relevant error variances and covariances or we might focus on the difference between estimated and observed values in actual extrapolations. In a linear system, the errors have their origins in system structure:

$$\hat{A}(L)y_t + \hat{B}(L)X_t = \hat{C}(L)e_t$$

The operator L is defined

$$L^j z_t = z_{t-j}$$

Estimated coefficients in the lag distribution matrices $\hat{A}(L)$, $\hat{B}(L)$, and $\hat{C}(L)$, are subject to sampling error

$$\text{var } \hat{A}(L), \text{ var } \hat{B}(L), \text{ var } \hat{C}(L);$$

and the residual error variance

$$\text{var } e_t$$

makes its additive contribution.

These error estimates are combined in a complicated nonlinear way for evaluation of standard error of forecast from

$$y_t = K\hat{\lambda}^t - [\hat{A}(L)]^{-1}\hat{B}(L)X_t + [\hat{A}(L)]^{-1}\hat{C}(L)e_t$$

λ^t is a vector of (powers of) eigenvalues of the system, and K is a matrix that depends on the initial conditions. Approximation formulas have been worked out for one-period solutions from such dynamic systems, but the numerical values for the formulas derived from replicated sampling experiments seem to be more satisfactory, especially in large systems.

A more synthetic approach is to look at the cumulative experience gained from repeated tabulations of

$$(y_t - \hat{y}_t)$$

These could refer to solutions from either linear or nonlinear systems. The underlying distribution theory for

$$\text{var } (y_t - \hat{y}_t)$$

is not known, but historical experience about the errors and error statistics has been accumulated, quarterly, over the past 14 years and shows some statistical stability. Some representative errors for extrapolations one year ahead are:

$$\Delta \text{GNP} \sim \$10 \text{ bill. or } \tfrac{2}{3}\%$$
$$\Delta \text{PGNP} \sim 1 \text{ index point or } \tfrac{1}{2}\%$$
$$\Delta U_n \sim \tfrac{1}{3} \text{ percentage point (unemployment)}$$
$$\Delta r_s \sim \tfrac{3}{4} \text{ index point (short-term interest rate)}$$
$$\Delta L \sim \tfrac{1}{2}\% \text{ (employment)}$$

These are not much larger than simulation errors over the sample period. Extrapolation errors are much larger, by a factor of 2 or 3.[5] But in realistic forward extrapolations, a great deal of extra, *a priori*, information is also taken into account. These numerical records of error experience are forecast errors from econometric models adjusted for *a priori* information.[6]

In the context of the American economy, these errors are not negligible, but they are tolerable – in the sense that useful economic policy decisions can be made subject to these errors. The regions of uncertainty that must surround policy choices because of these errors are not so large that they vitiate policy, but politicians cannot expect pinpoint precision, and they must be prepared to be wrong from time to time, perhaps as frequently as 20–30 per cent of the time. These are realistic limits.

One of the largest fields of application is in the comparison of policy alternatives. In this instance, we are comparing the difference (or ratio) between two solutions. Either parametric, model specification, or exogenous variable change will be introduced for such comparisons. There is good reason to believe that the comparison exercise will turn out to be more accurate than single solutions, without comparison. If there are common errors, either biases or similarities in distributions of random error will give rise to the possibility of cancellation of error effect.

Through model improvement (testing, respecifying, re-estimation, re-application) errors can be reduced a bit. This should be going on all the time. One strategy is to keep improving the description of the system through the model approximation. Another strategy is to use whatever system is available, and control or restrain it as much as possible in order to get some early results. The former seems to be more attractive at the moment, but that situation is subject to change.

CLASSES OF POLICY PROBLEMS

It is asserted that the degree of precision that has been attained so far is adequate to deal with the problems that policy-makers typically face. What are these problems? In the familiar area of demand management, the problems, the tools, and their application are quite straightforward. They consist of fiscal/monetary policies, and, in the case of the open economy, trade/payment policies. On a macro basis, the fiscal policies are amounts of public spending on current and capital account and the amount of taxes. Taxes may be approached through the rate structure, exemptions, deductions, and particular allowances—as in the case of incentives for capital formation. Monetary policy deals with the control of monetary aggregates such as $M_1, \ldots, M_7$ or interest rates, the representative US rate being the federal funds rate. Trade policy in the open economy consists of adjustment of schedules for duties, the promotion of exports, the promotion of import substitution, or manipulation of the exchange rate in a world of managed float. The awarding of public credits, grants, and aid use jointly fiscal and trade policies.

In the most straightforward management of economic policy, the authorities are assumed to be manipulating

$$
\begin{aligned}
G &= \text{public spending} \\
t &= \text{tax rate} \\
N &= \text{number of exemptions} \\
D &= \text{deductions from income} \\
M_i &= \text{monetary aggregate} \\
r_s &= \text{short term interest rate} \\
d &= \text{duty rate} \\
e_r &= \text{exchange rate.}
\end{aligned}
$$

This is a compact number of instruments and is suitable to a macro type model, around which the authorities can grasp the workings of the system as a whole. A limited number (5–10, at most) of instruments and policy targets are about as much as the non-professional human mind can properly comprehend and appreciate at one time.

The actual implementation of economic policy is rarely this simple, however. Politicians do not proceed with a systematic analysis of instruments and targets; they start immediately getting into details of:

Public spending for government jobs, for military, for transfers to local government, for public works, for specified subsidies.

Taxes for special incentives, sales taxes, property taxes, value added taxes, rebates, payroll allowances, social insurance taxes, luxury excises, vice-related excises.

Financial control over the designation of reserve rates (Eurodollar balances, time deposits, demand deposits), terms of credit contracts, stock market margins, public debt composition, permissible bank activities, interest rate ceilings.

Trade policy concerning specific duties (shoes, textiles, steel, TV sets), tax remission on exports, tax remission on non-repatriated earnings, acceptance of guest workers, sheltering of domestic-flag carriers in coastal trade, setting of quotas, granting of most favoured nation status.

Policy rarely works as we imagine it in textbooks. Legislators and administrators admire detail and tend to complicate what might otherwise be simplified rules. This means that even straightforward macro policies of the most pedestrian sort will get complicated for application, with many loopholes, shelters, and other special provisions. But this does not even lead us to the much more difficult route of reform and innovative policy.

Tax reform and welfare reform are two of the most pressing areas of economic management that need comprehensive policy changes. They will have substantial macro implications, but will be concerned with a vast number of minutiae. To some extent, these are innovative, but they do not begin to deal with the need for searching out of entirely new policies that will make the system perform better in both demand and supply respects. Policies to deal with the joint evils of inflation and unemployment, structural unemployment, investment in sensitive problem areas (energy and environment), monetary reform, trade liberalisation, and many similar problems are not going to be specified in terms of changes (along a one-dimensional scale) of exogenous variables or rates as listed above (G, t, N, D, M_i, r_s, d, e_r). These new and innovative policies are going to have to change and add entirely new equations to the accepted model. It is going to have to be a large integrated model of input-output, technology, sector price formation, components of national income, and components of national expenditure.

What I am saying is that the true and proper management of the economy does not proceed by manipulation of a selected number of exogenous variable instruments to guide the system towards a selected number of targets, subject to a fixed system of constraints. It is a vastly more complicated and more inelegant procedure, yielding less definitive conclusions than might be expected.

To make matters more precise, let us define the *control* problem for the economy, and examine the notional solution to this problem from the point of view of economic management as it must and is likely to be carried out. The model of the economy will be written as:

$$F(y'_t, \ldots, y'_{t-p}, Z'_t, \ldots, Z'_{t-p}, X'_t, \ldots, X'_{t-q}, w'_t, \ldots, w'_{t-q}, \theta') = e_t$$

and the policy decision function as:

$$L[(y_1 - y_1^*)', \ldots, (y_h - y_h^*)', X'_1, \ldots, X'_h] .$$

The deterministic *control* problem will be stated as:

$$\begin{aligned} &L = \text{min., subject to} \\ &x \\ &F = 0 . \end{aligned}$$

The meaning of this statement will be explained as follows:

y_t = endogenous target variables with lags up to order p.
X_t = exogenous instrument variables with lags up to order q.
Z_t = other endogenous variables, not subject to targeting, with lags up to order p.
w_t = other exogenous variables, not used as controlling instruments, with lags up to order q.
e_t = random errors.
θ = parameters of economic technical and statutory response.
h = length of policy horizon.

The function F is a column vector of relations with as many components as there are in y_t and Z_t together—as many equations in the economic model as endogenous variables. The function L is a loss function showing the cost of not meeting targets y_t^* together with the cost of instrumentation. The loss function is minimised over the policy planning horizon with respect to X, where each component of X must be assigned a value in each period 1, 2, . . ., h. The *deterministic* statement of the control problem is based on a point estimate of F; i.e., the value of the function when θ is assigned point values and e_t is placed at its mean.

$$\begin{aligned} \text{est } \theta &= \hat{\theta} \\ E(e_t) &= 0 \end{aligned}$$

Stochastic control would consist of choices for policy that produce performance within a region of the y_t-space, where the region is based on the probability distributions of θ and e_t, usually their variances

$$\text{var } \hat{\theta}$$
$$\text{var } e_t \, .$$

Given this statement of the problem, it is possible to review the various elements of policy control to see what is needed, what is uncertain, and what might be possible.

In order to implement control policy, it is first necessary to have an estimated model. This means having estimates, $\hat{\theta}$, of the parameters of F. Such models are available, and, naturally, I will speak in terms of my own experience with the Wharton Model. This model is frequently used for policy analysis and forecasting. It has gone through a number of validation tests. Having a model, we next choose targets y_t^*, instruments, X_t, and a policy horizon $1, 2, \ldots, h$. In a practical sense, y_t^* could be established for major aggregates such as full employment (4.5 per cent unemployment), high growth (potential GNP growing at 4 per cent from the last full employment date), trade balance (zero), low inflation (4 per cent change in the GNP deflator). Other macro targets might be set for domestic budget balance, investment ratio, wage share, interest rates, profit growth. The instruments to be used for achieving these targets are tax rates, public expenditure levels, transfer payments, monetary stock, short-term interest rate, tariff rate. Some numerical results have already been worked out for the Wharton Model with four targets and three instruments over a past horizon—17 quarters from 1971.1 to 1975.1. The computing problem is not essentially a barrier to this kind of model application, although it was once thought to be a serious obstacle. If the numbers of targets and instruments came to more than 25 each (or more) there could be problems of manageability, storage, and general size dimensions, but policy-makers are not able to make sense from so many magnitudes; therefore, we shall confine our attention to problems with small numbers of targets and instruments, as in the case of the application with the Wharton Model.

These calculations were made with a particular loss function

$$L = \sum_{t-1}^{h} \{ w_1(y_{1t} - y_{1t}^*)^2 + w_2(y_{2t} - y_{2t}^*)^2 + w_3(y_{3t} - y_{3t}^*)^2 + w_4(y_{4t} - y_{4t}^*)^2$$
$$+ \sigma_1 X_{1t}^2 + \sigma_2 X_{2t}^2 + \sigma_3 X_{3t}^2 \} \, .$$

This is a quadratic loss function with no cross-product terms. If L is quadratic and F is linear, the optimisation problem is especially simple and straightforward. If we relax the linearity of F, it is still not a very difficult computation and can be readily programmed to operate right off a general simulation program that has been independently constructed for the Wharton Model.

The results are evidently sensitive to specification of L. Within the framework of this parametric specification, we have experimented with alternative values for w_i and σ_i. If a target variable is to be effectively excluded, a high penalty can be attached to it. If an instrument is to be widely used, it can have a low penalty coefficient. In the contrary case, its effective use is curtailed.

Using high standards for employment, inflation, and growth, we find that some extreme values have to be chosen for X_t. This introduces *instrument instability*. One of the main reasons for including X_t in L is to penalise the use of instruments. This appeals to politicians, who are often pained if they feel compelled to use instruments vigorously. Instrument instability is a significant problem for control theory applications, and if the instruments are smoothed to realistic values, we find that we lose much on target approach.

This formulation of L is also weak in that no time discount factors are introduced. We could form a discounted stream of loss factors instead of taking a simple weighted mean with fixed weights. This would not add to computational difficulties, but it would require serious thought about the way to set the discount rate.

When L is minimised with respect to X_t, it is a case of finding 'best' values for these controlled exogenous variables. This is done, in the problem statement, subject to $\hat{F} = 0$, but economic poliy well be concerned with changing the constraint system as well as with searching the space of X_t for 'best' values. There is no *a priori* limit as to how F may be changed, but structural policies such as tax reform, welfare reform, commodity stabilisation, job training, environmental protection, conservation of materials, or energy expansion are likely (not necessarily) to replace components of F with new relationships. There is no formal method for searching for 'best' changes in F. It is a matter of surveying the situation, brainstorming innovative thinkers, and doing what is politically feasible. An incomes policy can potentially do more to bring inflation closer to its target than can any adjustment of X_t. In this sense, we do not obtain the minimum minimorum by following usual procedures.

Not only may F be changed deliberately in the economic policy process, but it may also fluctuate in a random way. Instead of using expected values for e_t, we could have random disturbances shift the F-function. From a computational standpoint, the disturbances should be drawn from the same (estimated) joint probability distribution that governed e_t in the sample period. At the same time, $\hat{\theta}$ is a random variable, dependent on sample data. Appropriate regions for joint variation of $\hat{\theta}$ and e_t can be constructed.[7] The formulas may be formidable to evaluate, especially in nonlinear systems, therefore numerical stochastic simulations, appropriately replicated, can be used instead to build numerically defined regions of uncertainty.

There are many caveats to this control theory construct, and the results lack uniqueness because of stochastic variation or change in system specification. Yet it is a powerful tool that is bound to be used on an increasing scale. It is interesting to examine how it worked out in applications with the Wharton Model in the turbulent period of the early 1970s. The objective of these calculations was to see whether 'optimal' policies could have improved the economic situation significantly. The policies that were actually used took the economy through a severe cyclical contraction, initiated by the oil embargo of autumn 1973, followed by OPEC pricing of crude oil at a fourfold multiple of its previous price.

At the end of the control period, real GNP would be 5 per cent higher under the control than by policies actually implemented. This appears to be a significant gain. It should be emphasised, however, that the recession of 1974–75 would not have been avoided. While the economy actually fell for five consecutive quarters, the 'optimal' path had two pauses on the way down, but the pattern would have qualified as a genuine recession. In the early part of the 17-quarter period, production would have been higher under the 'optimal' policies. It would then have sagged below actual values and recovered in late 1973 to out-perform the observed economy.

The price level (inflation rate) is practically the same for both cases—optimal control and actual. The unemployment rate would have looked a little better under 'optimal' control policies, but the gains are modest and do not begin to prevail systematically until 1974. The other target, for balanced trade, differed very little in the control mode from actual data. In some periods, the estimated balance is larger, while in others, the observed balance is to be preferred. Higher government spending and tax reduction are the fiscal tools used to improve the economy's performance. Spending at the beginning and end of the planning horizon is larger under the 'optimal' program. Unborrowed reserves, the monetary instrument, are intended to grow a bit faster under the control theory simulations. Finally, the tax rates (federal personal) are lower in most cases. Only the first bracket rate is shown here—for it can be used

TABLE 11.1a Optimal Control Solution of Wharton Model, 1971(1)–1975(1)

	Target Variables							
	Real GNP		*Unemployment Rate*		*Price Deflator*		*Trade Balance*	
	Optimal	*Actual*	*Optimal*	*Actual*	*Optimal*	*Actual*	*Optimal*	*Actual*
1971.1	743.1	736.9	5.90	5.95	139.3	139.5	2.3	2.9
2	750.4	742.1	5.81	5.91	140.9	141.1	−0.9	−0.1
3	753.4	747.2	5.83	5.96	141.9	142.0	−0.5	0.1
4	760.8	759.1	5.85	5.96	142.7	142.7	−3.5	−3.4
1972.1	776.8	770.9	5.70	5.82	144.6	144.6	−7.6	−7.1
2	787.7	786.6	5.58	5.68	145.5	145.3	−7.0	−6.9
3	795.7	798.1	5.52	5.57	146.7	146.5	−4.5	−4.8
4	807.7	814.2	5.33	5.31	148.3	148.0	−4.5	−5.3
1973.1	823.3	832.8	5.10	4.99	150.4	149.9	0.4	−0.8
2	830.3	837.4	5.05	4.91	152.9	152.6	1.3	0.5
3	839.3	840.8	4.89	4.76	155.8	155.7	6.8	6.8
4	850.2	845.7	4.81	4.75	158.7	158.9	8.4	9.3
1974.1	846.5	830.5	5.07	5.14	163.1	163.6	8.7	11.3
2	847.1	827.1	4.93	5.15	166.7	167.3	−4.6	−1.5
3	851.7	823.1	5.13	5.51	171.2	172.1	−7.7	−3.1
4	846.9	804.4	5.99	6.60	176.5	178.0	−5.1	1.8
1975.1	824.1	780.0	7.55	8.35	180.2	181.6	2.7	8.8

TABLE 11.1b Optimal Control Solution of Wharton Model 1971(1)–1975(1)

	Instrument Variables					
	Defence Spending		*Non-borrowed Reserves*		*Tax Rate* (1st *Bracket*)	
	Optimal	*Actual*	*Optimal*	*Actual*	*Optimal*	*Actual*
1971.1	76.7	71.8	30.0	29.7	13.4	14.0
2	77.2	70.8	29.7	29.8	13.3	14.0
3	74.5	70.0	29.4	29.9	13.8	14.0
4	72.5	72.1	30.3	30.8	14.3	14.0
1972.1	80.7	75.9	32.1	32.2	13.7	14.0
2	77.9	75.9	32.0	32.5	14.4	14.0
3	72.8	72.6	32.4	32.7	14.7	14.0
4	72.7	74.7	31.5	31.6	14.9	14.0
1973.1	70.6	75.0	31.1	30.7	14.9	14.0
2	71.6	74.0	31.2	30.4	14.4	14.0
3	73.5	73.3	32.7	31.8	13.6	14.0
4	76.2	75.3	34.8	33.5	13.1	14.0
1974.1	81.7	75.8	34.4	34.4	12.1	14.0
2	83.9	76.6	34.8	33.8	12.7	14.0
3	89.3	78.4	34.5	33.8	11.8	14.0
4	100.9	84.0	37.7	35.6	9.6	14.0
1975.1	104.6	84.7	36.1	35.7	10.9	14.0

as a simplified instrument although it is a good indicator for the whole rate structure.

The program is so designed that policy instruments do not get on widely changing paths, thus introducing the phenomenon of instrument instability. In our calculations, we penalised the use of instruments, so they would not be likely to move by unusually large swings that could not be accommodated.

There are some gains, but not spectacular ones. The 'optimal' economy is better, but the technique is not able to compensate fully for the very large disturbances and induced changes that actually occurred. In this sense, there is a genuine recession with high unemployment and two-digit inflation that cannot be eliminated.

While it could have done some good in the past in demand management, it does not seem to be the ideal tool to use in the supply side policies that are being considered for the future.

REASONABLE EXPECTATIONS AND FUTURE DEVELOPMENTS

It is wrong, misleading, and damaging to the cause of good economic management to argue in favour of the use of automatic and formal methods on the assumption that they are easy to manipulate; that they are quite precise; and that they are in a state of readiness. I have tried to convey quite a different impression, while still adhering strongly to the modelling approach.

A small inner core of a system or a condensed version can be used in easy manipulations of a limited number of exogenous variables. The results of these manipulations have very limited significance. They are broadly indicative of the main movements of the economy. They may often miss very important structural or exogenous variable changes because they are so small that they gloss over the strategic variables of a new situation.

It is, nevertheless, quite worth while and instructive to have the solution of the core system and its 'optimal' path. These will be indicative of what can be expected. The actual carrying out of policy will have to be more detailed, based on a much larger system that has all the statutory, behavioural, and technical reaction parameters in it. The 'optimal' solution of the core model can be regarded as a first approximation. The next step would be to build the supply-side system, to validate it, to use it for making relevant policy solutions–perhaps searching among a group of alternatives. The final step will be to make 'optimal' control simulations. These will be conditioned on the instruments and targets selected for the policy analysis. Here, it must be recognised that not more than ten or so targets with ten or so instruments can be conveniently analysed, either from the viewpoint of comprehending the joint meaning of all the targets or from the computational viewpoint. An absolute limit will be 25 at the present time, but the precise number is not known.

The numerical results discussed here and a good deal of the theory of optimal control of economic systems are based on deterministic calculation. It has been noted that stochastic control is more realistic. All the error terms in each structural equation of a system are set at mean values or some value based on *a priori* information. A better perspective of the ranges open to consideration of the degrees of uncertainty can be obtained from stochastic applications. It is not essential that stochastic control be implemented numerically. A great many theoretical properties have been investigated by analytic methods. It is probably the case, however, that in large systems of some hundreds of equations, it will be necessary to base applications on stochastic simulation methods. This kind of analysis will have, at least, two stochastic input channels, one associated with the estimated parameters of the system and the other with residual behavioural error tied to each equation (expect identities) of the system. The parameter estimates will have sampling error, and it is this error that gives rise to choice of alternative values for coefficients based on probability selection. The residual behavioural error does not fade away with increasing sample size. In this case, random numbers with given parametric characteristics, derived from sample period estimates of the covariance matrix will be used.

The ranges associated with stochastic optimal control simulations are analogous to tolerance intervals constructed with estimates of the standard error of forecast,

$$y_F \pm kS_{y_F},$$

where y_F is a forecast value, k is a scalar multiple (depending on probabilities of covering fractions of observations in the interval), and S_{y_F} is an estimate of the standard error of forecast. Experience gained from evaluation of standard errors of forecast indicate that the tolerance ranges grow quite rapidly, even for $k = 1$. It is my opinion that stochastic control simulations are going to provide fairly generous size regions in which there will be more latitude than policy-makers realize; nevertheless, they will not be so broad as to render the exercise pointless.

Apart from the numerical values generated in control theory applications – stochastic or deterministic – there are some qualitative results that can be assertained from theory. One such result is that in the face of uncertainty, there should be a tendency to reduce the amplitude of instrument change. Recommended policies should be steady. This is another manifestation of the instrument instability problem, that can be taken care of to some extent by introducing penalties to the use of instruments.

REFERENCES

1. Vijaya Duggal, L. R. Klein and M. D. McCarthy, 'The Wharton Model, Mark III: A Modern IS-LM Construct', *International Economic Review*, 15 (Oct 1974) 572–94.
2. L. R. Klein, 'Supply Constraints in Demand Oriented Systems: An Interpretation of the Oil Crisis', *Zeitschrift für Nationalokonomie*, 34 (1974) 45–56.
3. T. C. Liu and E.-C. Hwa, 'Structure and Applications of a Monthly Econometric Model of the US Economy', *International Economic Review*, xv (June 1974) 328–65.
4. See P. A. Samuelson, 'The Art and Science of Macromodels over 50 Years', *The Brookings Model: Perspective and Recent Developments*, ed. G. Fromm and L. R. Klein (Amsterdam: North Holland, 1975) pp. 2–10.
5. G. Fromm and L. R. Klein, 'The NBER/NSF Model Comparison Seminar: An Analysis of Results', *Annals of Economic and Social Measurement*, 5 (winter 1976) 1–28.
6. The representative errors are from S. McNees, 'An Evaluation of Economic Forecasts: Extension and Update', *New England Economic Review*, (Sep-Oct 1976) 30–44.
7. L. Johansen, 'Targets and Instruments under Uncertainty' (mimeo, 1975).

12 Economic Policy Formulation in a Large-scale Model

T. S. Barker

Methods of controlling econometric models have been developed mainly for small-scale linear models, usually for the purpose of demonstrating the power and potential application of control techniques. In the practice of economic management, however, the models used are often very large and nearly always nonlinear. This paper presents a method of formulating dynamic policies in a large-scale model, basically by adapting Tinbergen's targets-instruments approach (1963, 1966). The method is not intended to produce an optimal solution but rather a series of alternative solutions or strategies for economic policy, with different combinations of targets and instruments. The intention is to narrow down the feasible and sensible range of policies which are worth thorough and detailed exploration. The method is illustrated by an application to the Cambridge Growth Project's multisectoral dynamic model of the UK economy.

The paper is arranged as follows.

I A Multisectoral Dynamic Model. This section describes the large-scale, nonlinear model to which the method of formulating the policies is to be applied. Note however that the method is quite general in that it can be applied to any size of model.

II The Targets-Instruments Approach. This section develops the general approach and illustrates it in terms of two instruments and two targets over a 9-year planning horizon.

III An Application: increasing UK employment 1977–1985. The method is applied for the two instruments/targets example, solving the large-scale model for a progressive increase in employment of 100,000 a year.

IV Conclusions.

I A MULTISECTORAL DYNAMIC MODEL OF THE UK ECONOMY

The Cambridge Growth Project has developed a dynamic model from its medium-term static input-output model (Barker and others, 1978). The model follows the static version quite closely in form, with the main difference being that the equations and parameters in the dynamic model are intended to represent the year-to-year movements in the economy rather than the underlying trends. The main relationships and some dynamic multipliers for the first version of the model are given in (Barker, 1976). The current, second version (MDM2) has been used for this paper, but the form of the main relationships remains unchanged. The model has the following characteristics:

it is non-linear with a complex mixture of log-linear behavioural relationships, quantity and value accounting identities and systems of linear technical relationships such as input-output equations and classification converters;
it is large-scale with 2552 equations, 605 of which are stochastic; 7277 variables, 794 of which are policy instruments; and 11210 coefficients;
the main exogenous variables can be grouped into
policy instruments, fixed at current levels or announced projections, including the exchange rate
rest of the world variables such as competitors' export and domestic prices, industrial production by world area
North Sea oil and gas production
working population
average earnings per worker per year;
it is a medium-term model, based on annual time-series data and base-year (1970) cross-section data, projecting the economy year by year, 1973–85. Productivity and the growth in productive potential are *endogenous*, determined by the composition and growth of investment and the labour supply.

Thus it is clear that this model is of scale many times that of most quarterly econometric models. Methods of policy formulation have to be developed in the context of this scale and the costs of computing multiple solutions.

II THE TARGETS-INSTRUMENTS APPROACH

The final version of the static model made economic policies endogenous in the sense that the medium term could be projected assuming that the government maintains full employment and a balance of payments surplus. The method of doing this, based on the targets and instruments approach, is described by Livesey (1976).

This method can be generalised to the dynamic model by treating the n targets in t years as $n \times t$ targets, with $n \times t$ instruments for achieving them. Firstly, the linear relationships between each of the targets and all the instruments are measured by perturbing the model under the restriction that instruments can only affect targets in the same year or in subsequent years, not in earlier years. These linear relationships are then used as the basis for the solution of the instrument values required to achieve a given set of target values. Finally this solution is then inserted in the large-scale model and the procedure repeated until convergence. There are several variables in this procedure which can be changed in order to improve the algorithm: the size of the perturbation, the number of steps taken before a new linearisation is measured and the number of perturbations made for each linearisation. However in the application which follows the procedure has not been extensively tested in this way.

This approach is distinguished from that of Preston (1974) in having as many targets as there are time periods. Preston considers the problem of dynamic stabilisation with policy instruments changing each period to reach targets defined for one period in the future. The present approach emphasises the trade-off between values for target variables in different periods as well as that between targets and instruments.

In order to make the method clear, take an example with two instruments and two targets over the period 1977–85; this is the basis of the application to the large-scale model in the next section of the paper.

Definitions

EMPL	employment, thousands
BT	balance of trade, £ million 1970 export prices
SR	standard rate of income tax 1978 = 33
EX	exchange rate 1971 = 100

Linearised Condensed Form Equations

The targets are then estimated as linear functions of current and previous instrument values:

$$EMPL = \alpha_0 + \alpha_{10}SR + \alpha_{11}SR_{-1} + \alpha_{12}SR_{-2} + \ldots + \alpha_{20}EX + \alpha_{21}EX_{-1} + \alpha_{22}EX_{-2} + \ldots \quad (1)$$

$$BT = \beta_{10} + \beta_{10}SR + \beta_{11}SR_{-1} + \beta_{12}SR_{-2} + \ldots + \beta_{20}EX + \beta_{21}EX_{-1} + \beta_{22}EX_{-2} + \ldots$$

Targets and Instruments Solution

The targets, instruments and coefficients are 'stacked' in vectors and matrices to form:

$$\begin{bmatrix} EMPL85 \\ BT85 \\ EMPL84 \\ BT84 \\ \cdot \\ \cdot \\ \cdot \\ EMPL77 \\ BT77 \end{bmatrix} = \begin{bmatrix} \alpha_0 \\ \beta_0 \\ \alpha_0 \\ \beta_0 \\ \\ \\ \\ \alpha_0 \\ \beta_0 \end{bmatrix} \div \begin{bmatrix} \alpha_{10} & \alpha_{20} & \alpha_{11} & \alpha_{21} & \alpha_{12} & \alpha_{22} & \cdot & \cdot & \cdot & \cdot & \cdot & \cdot \\ \beta_{10} & \beta_{20} & \beta_{11} & \beta_{21} & \beta_{12} & \beta_{22} & \cdot & \cdot & \cdot & \cdot & \cdot & \cdot \\ 0 & 0 & \alpha_{10} & \alpha_{20} & \alpha_{11} & \alpha_{21} & & & & & & \\ 0 & 0 & \beta_{10} & \beta_{20} & \beta_{11} & \beta_{21} & \cdot & \cdot & \cdot & \cdot & \cdot & \cdot \\ \cdot & \cdot & \cdot & \cdot & \cdot & \cdot & & & & & & \\ \cdot & \cdot & \cdot & \cdot & \cdot & \cdot & & & & & & \\ \cdot & \cdot & \cdot & \cdot & \cdot & \cdot & & & & & & \\ 0 & 0 & 0 & 0 & 0 & 0 & \cdot & \cdot & \cdot & & \alpha_{10} & \alpha_{20} \\ 0 & 0 & 0 & 0 & 0 & 0 & \cdot & \cdot & \cdot & & \beta_{10} & \beta_{20} \end{bmatrix} \begin{bmatrix} SR \\ 85 \\ EX \\ 85 \\ SR \\ 84 \\ EX \\ 84 \\ \\ SR \\ 77 \\ EX \\ 77 \end{bmatrix}$$

$$+ \begin{bmatrix} & \vdots & \alpha_{19} & \alpha_{29} & \cdot & \cdot & \cdot \\ & \vdots & \beta_{19} & \beta_{29} & \cdot & \cdot & \cdot \\ & \vdots & \alpha_{18} & \alpha_{28} & \cdot & \cdot & \cdot \\ & \vdots & \beta_{18} & \beta_{28} & \cdot & \cdot & \cdot \\ 0 & \vdots & & & & & \\ & \vdots & \cdot & \cdot & & & \\ & \vdots & \cdot & \cdot & & & \\ & \vdots & \cdot & \cdot & & & \\ & \vdots & \alpha_{11} & \alpha_{21} & \cdot & \cdot & \cdot \\ & \vdots & \beta_{11} & \beta_{21} & \cdot & \cdot & \cdot \end{bmatrix} \begin{bmatrix} \\ 0 \\ \\ \hline SR76 \\ EX76 \\ \cdot \\ \cdot \\ \cdot \end{bmatrix} \qquad (3)$$

This can be characterised as

$$t_F = a + A\,i_F + B\,i_P \qquad (4)$$

where t is the vector of future targets, i_F is the vector of future

TABLE 12.1 Estimates of the condensed form equations for the targets in terms of the instruments

	Constant	Coefficients SR	SR_{-1}	SR_{-2}	SR_{-3}	SR_{-4}	EX	EX_{-1}	EX_{-2}	EX_{-3}	EX_{-4}	R^2	No. of observations
EMPL	α_0	α_{10}	α_{11}	α_{12}	α_{13}	α_{14}	α_{20}	α_{21}	α_{22}	α_{23}	α_{24}		
BT	β_0	β_{10}	β_{11}	β_{12}	β_{13}	β_{14}	β_{20}	β_{21}	β_{22}	β_{23}	β_{24}		
Model 1 (OLS)													
EMPL	0.1	−22.8	−1.6				−27.2	−10.2				0.994	9
	(0.1)	(4.4)	(3.2)				(11.2)	(2.3)					
BT	4.7	551.2	50.6				68.1	−91.1				0.978	9
	(1.7)	(102.8)	(76.0)				(26.5)	(55.0)					
Model 2 (OLS)													
EMPL	0.2	−31.3	−2.7				−27.2	−11.0				0.952	27
	(0.1)	(3.1)	(3.4)				(1.5)	(1.6)					
BT	2.7	500.2	−78.2				16.8	−135.4				0.893	27
	(1.3)	(43.5)	(47.5)				(21.2)	(22.3)					
Model 3 (OLS)													
EMPL	0.2	−48.1	−12.1	3.8	13.0		−46.5	−26.7	−11.0	−6.1	−29.8	0.966	45
	(0.5)	(9.6)	(13.0)	(14.1)	(11.6)		(5.8)	(6.6)	(6.9)	(7.6)	(7.4)		
BT	−0.2	114.4	−4.5	−32.6	29.2		12.0	−51.8	−43.6	−41.0	−93.3	0.952	45
	(1.2)	(21.1)	(28.6)	(31.1)	(25.6)		(12.9)	(14.6)	(15.1)	(16.9)	(16.3)		
Model 4 (OLS)													
EMPL	0.2	−48.5	0.7				−47.1	−23.8	−11.3	−5.1	−31.9	0.964	45
	(0.6)	(9.5)	(9.9)				(5.6)	(6.4)	(6.9)	(7.6)	(7.3)		
BT	−0.2	104.3					6.8	−46.5	−42.0	−43.0	−92.5	0.950	45
	(1.2)	(8.7)					(11.9)	(13.5)	(14.6)	(16.2)	(15.7)		

TABLE 12.1 (*continued*)

	Constant	Coefficients										R^2	No. of observations
		SR	SR_{-1}	SR_{-2}	SR_{-3}	SR_{-4}	EX	EX_{-1}	EX_{-2}	EX_{-3}	EX_{-4}		
Model 5 (direct estimates for changes MDM2)													
EMPL	0	−38.5	−4.2	−1.8	2.4	1.8 . . .	−35.4	−21.7	2.7	4.1	3.7 . . .		
BT	0	110.2	−26.8	25.5	−24.5	2.8 . . .	9.3	−61.0	−36.4	−25.8	−17.9 . . .		

Notes: (1) Numbers in brackets are standard errors,
(2) Coefficients refer to SR(1978 = 33) and EX(1971 = 100).

instruments and i_P the vector of past instruments. The solution for the instrument vector is

$$i_F = A^{-1}[t_F - a - B\,i_P] \tag{5}$$

Note that the coefficients in equations (1) and (2) are assumed to be time-invariant, so that they are repeated for each pair of targets in each year of the planning period in equation (3). If information was available on any change in the coefficients from year to year, it could be used in equation (3). Note also that if the values of the targets and instruments are transformed into *changes* in values from a standard projection then the last term in equations (3) and (4) becomes zero because the past values of instruments remain unchanged. With the additional restriction of no constants in equations (1) and (2), equation (5) becomes

$$i_F = A^{-1}\,t_F \tag{6}$$

III AN APPLICATION: INCREASING UK EMPLOYMENT, 1977–85

The application of the method of section II above mainly involves the estimation of the matrix A containing the coefficients from linear relationships between targets and instruments. Two types of estimate have been made, both involving *changes* in the target and instrument values from a particular projection of the economy using the model. The first type is obtained by OLS regression of the changes in target values on the changes in instrument values obtained by perturbing an initial projection (the 'no policy change' standard view of the future economy). The perturbations were an ordered sequence of $\pm$ 10 per cent in the instrument values. The results of 4 sets of regressions (models 1 to 4) are shown in Table 12.1. The second type of estimate is the direct measurement of the effects of a sustained change in each instrument over the planning period. These results are shown as model 5 in Table 12.1. They are derived from the results given in Tables 12.2 and 3 which show the differences from standard for the changes and the cumulation of differences respectively.

The OLS results proved to be too unstable when the perturbations were changed. In addition it seemed better to give the effects of changing the instruments from the first year of the planning period much greater weight than changes in other years. Thus the direct estimates were used to form the coefficients of matrix A given in Table 12.4 with the inverse in Table 12.5.

TABLE 12.2 Changes in employment and the balance of trade following changes in three key policy instruments: MDM2

difference from standard

Year	*Employment (thousands)*		*Balance of trade (£ million 1970 export prices)*	
	−10% in exchange rate	*−10% in income tax rates*	*−10% in exchange rate*	*−10% in income tax rates*
1977	248	129	−65.4	−369.0
1978	400	143	361.4	−279.3
1979	381	149	616.5	−364.7
1980	352	141	794.4	−282.7
1981	326	135	919.9	−292.1
1982	318	132	944.9	−293.7
1983	334	130	392.9	−308.7
1984	352	131	866.1	−356.8
1985	363	130	753.3	−357.0

TABLE 12.3 Changes in employment and the balance of trade following changes in three-key policy instruments: cumulative differences MDM2

difference from standard

Year	*Employment (thousands)*		*Balance of trade (£ million 1970 export prices)*	
	−10% in exchange rate	*−10% in income tax rates*	*−10% in exchange rate*	*−10% in income tax rates*
1977 over 76	248	129	−65.4	−369.0
1978 – 77	152	14	426.8	89.7
1979 – 78	−19	6	255.1	−85.4
1980 – 79	−29	−8	177.9	82.0
1981 – 80	−26	−6	125.5	−9.4
1982 – 81	−8	−3	25.0	−1.6
1983 – 82	16	−2	−52.0	−15.0
1984 – 83	18	1	−26.8	−48.1
1985 – 84	11	−1	−112.8	−0.2

The results in Table 12.5 are best explained by an analysis of the solutions of the instrument values which can be derived for different patterns of required changes in the targets. Six solutions are given in Table 12.6 for three types of change in the two targets. Thus in order to produce an extra 100,000 in employment in 1977 returning to the

TABLE 12.4 Matrix *A*: direct estimates MDM2, 1977–85

Changes in	*SR*85	*EX*85	*SR*84	*EX*84	...												*SR*77	*EX*77
EMPL85	−38.5	−35.4	−4.2	−21.7	−1.8	2.7	2.4	4.1	1.8	3.7	0.9	1.1	0.6	−2.3	−0.3	−2.8	0.3	−1.6
BT85	110.2	9.3	−26.8	−61.0	25.5	−36.4	−24.5	−25.8	2.8	−17.9	0.5	−3.6	4.5	7.3	14.4	3.8	0.1	16.1
EMPL84	0.0	0.0	−38.5	−35.4	−4.2	−21.7	−1.8	2.7	2.4	4.1	1.8	3.7	0.9	1.1	0.6	−2.3	−0.3	−2.8
BT84	0.0	0.0	110.2	9.3	−26.8	−61.0	25.5	−36.4	−24.5	−25.8	2.8	−17.9	0.5	−3.6	4.5	7.3	14.4	3.8
.	0.0	0.0	0.0	0.0	−38.5	−35.4	−4.2	−21.7	−1.8	2.7	2.4	4.1	1.8	3.7	0.9	1.1	0.6	−2.3
.	0.0	0.0	0.0	0.0	110.2	9.3	−26.8	−61.0	25.5	−36.4	−24.5	−25.8	2.8	−17.9	0.5	−3.6	4.5	7.3
.	0.0	0.0	0.0	0.0	0.0	0.0	−38.5	−35.4	−4.2	−21.7	−1.8	2.7	2.4	4.1	1.8	3.7	0.9	1.1
	0.0	0.0	0.0	0.0	0.0	0.0	110.2	9.3	−26.8	−61.0	25.5	−36.4	−24.5	−25.8	2.8	−17.9	0.5	−3.6
	0.0	0.0	0.0	0.0	0.0	0.0	0.0	0.0	−38.5	−35.4	−4.2	−21.7	−1.8	2.7	2.4	4.1	1.8	3.7
	0.0	0.0	0.0	0.0	0.0	0.0	0.0	0.0	110.2	9.3	−26.8	−61.0	25.5	−36.4	−24.5	−25.8	2.8	−17.9
	0.0	0.0	0.0	0.0	0.0	0.0	0.0	0.0	0.0	0.0	−38.5	−35.4	−4.2	−21.7	−1.8	2.7	2.4	4.1
	0.0	0.0	0.0	0.0	0.0	0.0	0.0	0.0	0.0	0.0	110.2	9.3	−26.8	−61.0	25.5	−36.4	−24.5	−25.8
	0.0	0.0	0.0	0.0	0.0	0.0	0.0	0.0	0.0	0.0	0.0	0.0	−38.5	−35.4	−4.2	−21.7	−1.8	2.7
	0.0	0.0	0.0	0.0	0.0	0.0	0.0	0.0	0.0	0.0	0.0	0.0	110.2	9.3	−26.8	−61.0	25.5	−36.4
	0.0	0.0	0.0	0.0	0.0	0.0	0.0	0.0	0.0	0.0	0.0	0.0	0.0	0.0	−38.5	−35.4	−4.2	−21.7
	0.0	0.0	0.0	0.0	0.0	0.0	0.0	0.0	0.0	0.0	0.0	0.0	0.0	0.0	110.2	9.3	−26.8	−61.0
EMPL77	0.0	0.0	0.0	0.0	0.0	0.0	0.0	0.0	0.0	0.0	0.0	0.0	0.0	0.0	0.0	0.0	−38.5	−35.4
BT77	0.0	0.0	0.0	0.0	0.0	0.0	0.0	0.0	0.0	0.0	0.0	0.0	0.0	0.0	0.0	0.0	110.2	9.3

TABLE 12.5 Matrix A^{-1}: direct estimates MDM2, 1977–85

Changes in	*EMPL85*	*BT85*	*EMPL84*	*BT84*	...												*EMPL77*	*BT77*
SR85	2.6	10.0	−20.0	−4.5	9.6	−0.7	−8.5	0.1	2.8	0.1	−1.9	−0.7	4.8	1.3	−4.1	−1.8	10.5	3.2
EX85	−31.1	−10.9	40.4	10.3	−35.4	−6.3	30.5	4.3	−23.6	−3.0	17.8	2.8	−15.2	−2.4	13.5	3.4	−19.3	−5.3
SR84	0.0	0.0	2.6	10.0	−20.0	−4.5	9.6	−0.7	−8.5	0.1	2.8	0.1	−1.9	−0.7	4.8	1.3	−4.1	−1.8
EX84	0.0	0.0	−31.1	−10.9	40.4	10.3	−35.4	−6.3	30.5	4.3	−23.6	−3.0	17.8	2.8	−15.2	−2.4	13.5	3.4
.	0.0	0.0	0.0	0.0	2.6	10.0	−20.0	−4.5	9.6	−0.7	−8.5	0.1	2.8	0.1	−1.9	−0.7	4.8	1.3
.	0.0	0.0	0.0	0.0	−31.1	−10.9	40.4	10.3	−35.4	−6.3	30.5	4.3	−23.6	−3.0	17.8	2.8	−15.2	−2.4
.	0.0	0.0	0.0	0.0	0.0	0.0	2.6	10.0	−20.0	−4.5	9.6	−0.7	−8.5	0.1	2.8	0.1	−1.9	−0.7
	0.0	0.0	0.0	0.0	0.0	0.0	−31.1	−10.9	40.4	10.3	−35.4	−6.3	30.5	4.3	−23.6	−3.0	17.8	2.8
	0.0	0.0	0.0	0.0	0.0	0.0	0.0	0.0	2.6	10.0	−20.0	−4.5	9.6	−0.7	−8.5	0.1	2.8	0.1
	0.0	0.0	0.0	0.0	0.0	0.0	0.0	0.0	−31.1	−10.9	40.4	10.3	−35.4	−6.3	30.5	4.3	−23.6	−3.0
	0.0	0.0	0.0	0.0	0.0	0.0	0.0	0.0	0.0	0.0	2.6	10.0	−20.0	−4.5	9.6	−0.7	−8.5	0.1
	0.0	0.0	0.0	0.0	0.0	0.0	0.0	0.0	0.0	0.0	−31.1	−10.9	40.4	10.3	−35.4	−6.3	30.5	4.3
	0.0	0.0	0.0	0.0	0.0	0.0	0.0	0.0	0.0	0.0	0.0	0.0	2.6	10.0	−20.0	−4.5	9.6	−0.7
	0.0	0.0	0.0	0.0	0.0	0.0	0.0	0.0	0.0	0.0	0.0	0.0	−31.1	−10.9	40.4	10.3	−35.4	−6.3
	0.0	0.0	0.0	0.0	0.0	0.0	0.0	0.0	0.0	0.0	0.0	0.0	0.0	0.0	2.6	10.0	−20.0	−4.5
	0.0	0.0	0.0	0.0	0.0	0.0	0.0	0.0	0.0	0.0	0.0	0.0	0.0	0.0	−31.1	−10.9	40.4	10.3
SR77	0.0	0.0	0.0	0.0	0.0	0.0	0.0	0.0	0.0	0.0	0.0	0.0	0.0	0.0	0.0	0.0	2.6	10.0
EX77	0.0	0.0	0.0	0.0	0.0	0.0	0.0	0.0	0.0	0.0	0.0	0.0	0.0	0.0	0.0	0.0	−31.1	−10.9

Note: coefficients have been scaled by 1000.

TABLE 12.6 An analysis of a solution of two instruments to changes in two targets

	The employment target								
	Change in EMPL	*Required changes in SR*	*EX*	*Change in EMPL*	*Required changes in SR*	*EX*	*Change in EMPL*	*Required changes in SR*	*EX*
1977	100	0.3	−3.1	100	0.3	−3.1	100	0.3	−3.1
78	0	−2.0	4.0	100	−1.7	0.9	200	−1.5	−2.2
79	0	1.0	−3.5	100	−0.8	−2.6	300	−2.2	−4.8
80	0	−0.9	3.0	100	−1.6	0.4	400	−3.9	−4.3
81	0	0.3	−2.4	100	−1.3	−1.9	500	−5.2	−6.2
82	0	−0.2	1.8	100	−1.5	−0.1	600	−6.8	−6.4
83	0	0.5	−1.5	100	−1.1	−1.7	700	−7.8	−8.0
84	0	−0.4	1.3	100	−1.5	−0.3	800	−9.3	−8.4
85	0	1.0	−1.9	100	−0.4	−2.2	900	−9.7	−10.6

	The balance of trade target								
	Change in BT	*Required change in SR*	*EX*	*Change in BT*	*Required change in SR*	*EX*	*Change in BT*	*Required change in SR*	*EX*
1977	100	1.0	−1.1	100	1.0	−1.1	100	1.0	−1.1
78	0	−0.4	1.0	100	0.6	−0.1	200	1.6	−1.1
79	0	−0.1	−0.6	100	0.5	−0.7	300	2.0	−1.8
80	0	0.0	0.4	100	0.5	−0.3	400	2.5	−2.1
81	0	0.0	−0.3	100	0.5	−0.6	500	3.0	−2.6
82	0	−0.1	0.3	100	0.4	−0.3	600	3.5	−2.9
83	0	0.1	−0.2	100	0.6	−0.5	700	4.0	−3.0
84	0	−0.2	0.3	100	0.4	−0.2	800	4.4	−3.6
85	0	0.3	−0.5	100	0.7	−0.7	900	5.1	−4.3

Notes: (1) Definitions
EMPL total employment, thousands
BT balance of trade, £ million 1970 export prices
SR standard rate of tax 1978 = 33
EX exchange rate 1971 = 100

(2) The changes are measured from the direct estimates of the condensed form, 10% change, MDM2 no policy change.

original projections thereafter, the exchange rate must fall by 3.1 points in 1977 whilst the tax rate increases by 0.3 points in order to nullify the balance of trade deficit which would otherwise emerge from the depreciation. The changes are reversed in direction in 1978 and become larger in magnitude to offset the lagged effects of the 1977 changes as well as bring employment back to the standard view. The required changes in instruments oscillate throughout the planning period, diminishing slowly in magnitude. This is a dramatic illustration of the need for policy makers to take the medium-term consequences of their current policies into account.

The steady increase in employment, giving the third set of required

TABLE 12.7 Full model solutions for a gradual increase in employment of 100,000 a year 1977–85

absolute differences from standard

	1977	1978	1979	1980	1981	1982	1983	1984	1985
Desired solution									
EMPL	100	200	300	400	500	600	700	800	900
BT	0	0	0	0	0	0	0	0	0
Preferred method									
First solution									
SR	0.3	−1.5	−2.2	−3.9	−5.2	−6.8	−7.8	−9.3	−9.7
EX	−3.1	−2.2	−4.8	−4.3	−6.2	−6.4	−8.0	−8.4	−10.6
EMPL	96	189	270	339	408	475	544	589	674
BT	4	−25	17	16	29	−73	−174	327	−554
Second solution									
SR	0.2	−1.3	−2.6	−4.3	−6.2	−7.0	−8.1	−8.8	−8.8
EX	−3.2	−2.8	−4.9	−5.5	−7.1	−9.1	−10.7	−13.7	−15.5
EMPL	100	203	308	390	479	569	655	794	884
BT	−1	—	—	44	−5	83	−19	−61	−165
Third solution									
SR	0.2	−1.3	−2.6	−4.6	−6.2	−8.1	−7.9	−9.0	−7.3
EX	−3.2	−2.7	−4.8	−5.5	−7.6	−8.5	−12.4	−12.6	−18.5
EMPL	100	200	302	397	500	604	734	773	941
BT	−1	—	—	3	18	−77	9	81	14
Fourth solution									
SR	0.2	−1.3	−2.6	−4.6	−6.4	−7.2	−8.5	−9.0	−8.0
EX	−3.2	−2.7	−4.8	−5.6	−7.3	−9.7	−10.2	−14.7	−15.0
EMPL	100	201	301	399	496	596	661	811	873
BT	−1	—	−1	5	−12	47	−47	−107	13
Alternative method									
First solution as above									
Second solution									
SR	0.2	−1.3	−2.6	−4.3	−6.2	−8.1	−8.1	−8.7	−8.8
EX	−3.2	−2.8	−4.9	−5.6	−6.8	−9.4	−10.2	−14.3	−15.2
EMPL	100	204	308	393	473	578	646	814	881
BT	−1	−1	2	46	−5	27	−7	−81	−111

Note: for definitions see Table 12.6.

changes in instrument values (shown in the top panel of Table 12.6) has been taken as the target for the large-scale model. The instrument changes have been fed back into the model and the effects on the targets are given as the first solution in Table 12.7. The approximation is very close for the first year but deteriorates thereafter, particularly after 1982. The employment changes are consistently less than required whilst the balance of trade veers sharply into deficit after 1982 instead of remaining unchanged at the standard view levels. One reason for the deficit seems to be the interaction of the lower exchange rate and the lower tax rates: a reduction in the tax rate has a greater effect on the balance of trade when the exchange rate is lower because import prices are higher. There are at least two other reasons for the discrepancies: the assumption of linearity in the responses and the assumption that the estimates measured for changes in 1977 apply equally throughout the period.

The changes required to move from the first solution to the desired solution of the targets are then used to form a new set of implied instrument changes. The matrix A is left unchanged. The second, third and fourth solutions by this method are also given in Table 12.7. Since we are primarily interested in the first few years of the solution, it is not necessary to make any more solutions.

An alternative method was also tried – this time re-estimating matrix A after each solution of the full model. Thus each solution in this method requires three runs of the model. The second solution, equivalent in computing costs to the fourth solution of the preferred method, is shown in Table 12.7 and is clearly inferior.

IV CONCLUSIONS

This paper has demonstrated the feasibility of applying the targets-instruments approach to a large-scale nonlinear model. The method of doing so is quite general in that it can be applied to any model of the economy and to any combination of linearly-independent targets and linearly-independent instruments for achieving them. The procedure, as described, is however limited in two respects. First, it achieves targets year by year, rather than over a number of years (for example it may be better to define the balance of trade target in terms of a required surplus over the whole planning period rather than for each year separately). Second, it can yield infeasible solutions for the instruments both for their levels and for their paths. This last limitation means that the targets may have to be revised in the light of their implications for the instrument values. However such an exercise is valuable in emphasising how interdependent the targets and instruments are in policy formulation.

The paper has also given a dramatic illustration of the arguments for formulating policies with regard to the medium term as opposed to the short term. Not only have many policies to be planned well in advance of their implementation, but they also have effects which continue for years afterwards. If these consequences are not taken into account, then medium-term targets will not be achieved because it is only feasible to change instruments gradually. Furthermore the consequences usually imply future changes in instruments to bring the economy back on target and if the changes are unacceptable the targets will not be reached.

REFERENCES

T. S. Barker (1976) (ed.), *Economic Structure and Policy* (Chapman & Hall, London).

T. S. Barker (1978), 'Towards strategic paths in economic planning in J. R. N. Stone and W. Peterson, (ed.), *Econometric Contributions to Public Policy* (Macmillan, London).

D. A. Livesey (1976), 'Solving the model' Chapter 3 in Barker, *Economic Structure and Policy*, above.

A. J. Preston (1974), 'A dynamic generalisation of Tinbergen's theory of policy', *Review of Economic Studies*, no. 125 (Jan) pp. 65–74.

J. Tinbergen (1963), *On the Theory of Economic Policy* (North Holland, 2nd ed.).

J. Tinbergen (1966), *Economic Policy: Principles and Design* (North Holland).

Index